MINOUX
Mathematical Programming: Theory and Algorithms
(*Translated by S. Vajda*)

MIRCHANDANI AND FRANCIS, EDITORS
Discrete Location Theory

NEMHAUSER AND WOLSEY
Integer and Combinatorial Optimization

NEMIROVSKY AND YUDIN
Problem Complexity and Method Efficiency in Optimization
(*Translated by E. R. Dawson*)

PALMER
Graphical Evolution: An Introduction to the Theory
of Random Graphs

PLESS
Introduction to the Theory of Error-Correcting Codes
Second Edition

SCHRIJVER
Theory of Linear and Integer Programming

TOMESCU
Problems in Combinatorics and Graph Theory
(*Translated by R. A. Melter*)

TUCKER
Applied Combinatorics
Second Edition

Probabilistic Analysis of Packing and Partitioning Algorithms

Probabilistic Analysis of Packing and Partitioning Algorithms

E. G. COFFMAN, JR.
AT&T Bell Laboratories
Murray Hill, New Jersey

GEORGE S. LUEKER
Department of Information and Computer Science
University of California
Irvine, California

A Wiley-Interscience Publication
JOHN WILEY & SONS, INC.
New York · Chichester · Brisbane · Toronto · Singapore

In recognition of the importance of preserving what has been
written, it is a policy of John Wiley & Sons, Inc., to have books
of enduring value published in the United States printed on
acid-free paper, and we exert our best efforts to that end.

Copyright © 1991 by AT&T.

All rights reserved. Published simultaneously in Canada.

Reproduction or translation of any part of this work
beyond that permitted by Section 107 or 108 of the
1976 United States Copyright Act without the permission
of the copyright owner is unlawful. Requests for
permission or further information should be addressed to
the Permissions Department, John Wiley & Sons, Inc.

Library of Congress Cataloging-in-Publication Data:
Coffman, E. G. (Edward Grady), 1934-
   Probabilistic analysis of packing and partitioning algorithms/
E.G. Coffman, Jr., George S. Lueker.
     p.   cm. — (Wiley-Interscience series in discrete mathematics
and optimization)

   "A Wiley-Interscience publication."
   Includes bibliographical references and index.
   ISBN 0-471-53272-X
   1. Partitions (Mathematics). 2. Combinatorial packing and
covering. 3. Probabilities. I. Lueker, George S. II. Title.
III. Series.

QA165.C64   1991
512.9'25—dc20
                                       90-23539
                                             CIP

Printed in the United States of America

10 9 8 7 6 5 4 3 2 1

To our parents ·

Preface

In the early 1970's the development of the theory of NP-completeness gave strong evidence that, for many optimization problems, it would be unrealistic to hope to find efficient algorithms that always produce the exact optimum. This provided new motivation for the study of algorithms that give good approximations, and for algorithms that work well on the average. Since the mid-1970's, a great deal of research has concentrated on the application of probability theory to the analysis of algorithms. This book examines techniques that have proven useful in such analysis, focusing on applications to two problem areas: bin packing and partitioning. Since the book concentrates as much on techniques as on results, it should provide a useful introduction to probabilistic analysis even for those readers with interests in other problem areas.

In a typical packing problem, we want to pack a set of objects into standardized containers so as to minimize wasted space; equivalently, we may want to cut a required set of pieces from standardized stock so as to minimize wasted material (trim loss). In the scheduling problem, we want to allocate a collection of jobs (without precedence constraints) among a set of machines so as to minimize the time until all jobs are finished. This scheduling problem and the one-dimensional versions of the packing and cutting problems can all be described abstractly in terms of set partitioning under sum constraints. One or more of these problems, or their generalizations to two or three dimensions, exist in virtually every modern industry; collectively, they are of great practical importance. All currently known algorithms for finding exact solutions to these problems require a computing time that becomes impractical as the number of items (objects, pieces, jobs) becomes large; the theory of NP-completeness suggests that this will continue to be the case. Thus, a major effort has gone into the design and analysis of fast heuristic algorithms that produce good, approximate solutions most of the time. A mathematical analysis is clearly needed in order to give precise meanings to the terms "good" and "most of the time." Such an analysis can also be expected to yield insights into the structure of good heuristics.

This book presents and illustrates a wide variety of techniques for analyzing the typical or average-case properties of the solutions (packings, cuttings, or schedules) produced by heuristic algorithms. In terms of the packing problem, for example, the analysis begins with the formulation of a probability model in which the items to be packed are defined to be random samples from some given probability distribution. Properties of the packings produced by the algorithm then become random variables whose distributions are the goals of the analysis. Intensive research into the analysis of this problem began a little over 10 years ago, and picked up considerable momentum a few years later with striking new results.

Major mathematical challenges are posed by the probabilistic analysis of the problems presented here. Indeed, as the reader will soon discover, approaches leading to exact results are rarely successful, especially for the better heuristics. Instead, researchers have turned to asymptotic methods, a secondary theme of this book. Such methods provide order-of-magnitude results for problem instances containing large numbers of items. In spite of their limitations, the results can be very useful in evaluating and comparing heuristics.

This book grew out of a survey on asymptotic methods that we wrote in collaboration with A. H. G. Rinnooy Kan [CLR88]. It is a pleasure to acknowledge the part played by Rinnooy Kan in originating the survey and in subsequently encouraging us to follow up with a book. We have benefitted greatly from the discussions we have had with many of our colleagues. Notable among these are Leo Flatto, David S. Johnson, Richard Karp, Kadri Krause, WanSoo Rhee, Peter Shor, Michel Talagrand, and Richard Weber. We owe a special debt of gratitude to Peter Shor for his contributions, particularly in collaborating with us in the preparation of Chapter 3 and in granting permission for us to adapt material from [Shor86]. Finally, we would like to thank Sue Pope for her expert typing assistance, and Pat Harris for her skillful assistance in preparing the files necessary to set the manuscript using LaTeX. The second author wishes to thank the National Science Foundation[1] for support under Grants DCR 85-09667 and CCR 89-12063 at the University of California at Irvine.

---

[1] Any opinions, findings, and conclusions or recommendations expressed in this publication are those of the authors and do not necessarily reflect the views of the National Science Foundation.

Contents

List of Figures   xiii

1 Introduction   1
   1.1 Overview 1
   1.2 Illustrative applications 4
   1.3 Notation 5
   1.4 Classical algorithms 7
       1.4.1 Makespan scheduling 7
       1.4.2 Bin packing 8

2 Analysis Techniques   11
   2.1 Sums of i.i.d. random variables 12
       2.1.1 Small deviations and the central limit theorem 12
       2.1.2 Bounds on the tails of the distributions 15
       2.1.3 Estimates of moments 22
   2.2 Markov chains 23
   2.3 Bounds 25
   2.4 Dominating algorithms 26
   2.5 Bounds that usually hold 27
   2.6 Monotonicity 30
   2.7 More specialized techniques 32
       2.7.1 Applications of the Poisson process 32
       2.7.2 Kolmogorov-Smirnov statistics 34
       2.7.3 The second moment method 37
       2.7.4 An application of renewal theory 38

3 Matching problems — 41
3.1 Proofs for Euclidean and rightward matching — 43
3.1.1 The lower bound — 44
3.1.2 The upper bound — 50
3.1.3 A rightward matching problem — 53
3.2 Proof of the up-right matching estimate — 56
3.2.1 The lower bound — 57
3.2.2 The upper bound — 64

4 Scheduling and Partitioning — 75
4.1 Analysis of classical greedy heuristics — 75
4.2 Differencing methods — 83
4.3 On the optimum solution — 90

5 Bin Packing: The Optimum Solution — 99
5.1 Basic algorithms and bounds — 99
5.2 Perfect packings — 104
5.3 Functional analysis of the packing constant — 110

6 Bin Packing: Heuristics — 121
6.1 Off-line packing: FFD and BFD — 122
6.1.1 The expected behavior — 122
6.1.2 Deviation from the expected behavior — 128
6.2 On-line bin packing: Best Fit — 129
6.3 On-line linear-time bin packing — 133
6.3.1 Next Fit: The expected behavior — 133
6.3.2 Deviation from the expected behavior — 144
6.3.3 The HARMONIC algorithm — 146
6.3.4 On-line matching — 148
6.3.5 On-line packing with limited active bins — 150

7 Packings in Two Dimensions — 155
7.1 Off-line algorithms — 155
7.1.1 Packing squares into a strip — 156
7.1.2 Packing rectangles into a strip — 166
7.1.3 Two-dimensional bin packing — 169
7.2 On-line algorithms — 172

CONTENTS

References 177

Index 189

List of Figures

2.1	Illustration of the argument used in the proof of Hoeffding's bound	18
2.2	Packing about a reciprocal	29
3.1	A maximum up-right matching	42
3.2	A square S and its triangular regions	45
3.3	The pairs in S_{ij}	47
3.4	Moving plus points of $S \in \mathcal{G}_{i-1}$ in stage 1, step i	52
3.5	A rightward matching	54
3.6	The transformed problem	58
3.7	The sequence of triangles after a stage of the refinement process	59
3.8	The initial triangle	60
3.9	A typical triangle T	61
3.10	Illustration of the region R	65
3.11	Approximating a function in $\mathcal{F}$ by a function in $\mathcal{F}^*$	67
4.1	The function $u(x)$ and two inverse functions, α and β	78
4.2	Illustration of PDM	84
4.3	Illustration of LDM	85
4.4	Sampling to control distributions	88
5.1	An example of a dual-feasible function	108
5.2	Bin packing with items drawn uniformly from $[a,b]$	109
5.3	Illustration of a simple proof that $U(a,b)$ allows perfect packing if $[a,b] \subseteq [0,1]$ is symmetric about $1/p$	115
5.4	A perfect packing strategy for the interval $[a,b]$	117
6.1	Example of the queueing process for MFFD	125
6.2	An example of Best Fit	130

6.3	An example of the matching used during the proof of the upper bound on Best Fit	131
6.4	An example of the matching used during the proof of the lower bound for open-end on-line packing	134
6.5	Partition examples for the analysis of Next Fit	139
6.6	The mapping cycle	140
6.7	The asymptotic efficiency of NF packings	145
6.8	An example for the HARMONIC algorithm	147
6.9	Illustration of the argument leading to an integral equation for Smart Next Fit	152
6.10	Comparison of the efficiency of Next Fit and Smart Next Fit	153
7.1	A strip-packing example	156
7.2	A square-packing example	157
7.3	Example for the first three steps of Algorithm A	160
7.4	The function $\delta(y)$	162
7.5	A packing produced by Algorithm B	167
7.6	Example for Algorithm KLM	170
7.7	A BFS packing	176

Probabilistic Analysis
of Packing and
Partitioning Algorithms

Chapter 1

Introduction

1.1 Overview

The problems to be studied in this book generally require the partitioning of a set $S = \{X_1, X_2, \ldots, X_n\}$ of nonnegative numbers so that the sums of the elements in the blocks of the partition satisfy some given property. For example, an instance of the *makespan scheduling problem* is made up of S and an integer $m \geq 2$; the objective is to find a partition of S into m blocks so that the maximum block sum (i.e., the makespan) is minimized over all such partitions. Although there are many possible interpretations, in the one motivating the name of the problem each X_i represents a task or job running time and a block corresponds to the set of tasks to be run on the same processor of an m-processor multiprocessor system.

Very closely related is the *partition problem*. The input for this problem is the same as for the makespan scheduling problem, but now the objective is to find a partition of L_n into m blocks such that the difference between the maximum and minimum block sums is minimized over all such partitions. Note that for the special case $m = 2$ an optimum partition (schedule) for the partition problem is also an optimum schedule for the makespan problem. Moreover, for any $m \geq 2$ one expects that a good heuristic for the partition problem will be a good heuristic for makespan scheduling.

As another example, an instance of the *one-dimensional bin-packing problem* is composed of a real number $c > 0$ (the bin capacity) and a set S of elements (items or pieces) no larger than c. The problem is to find a partition of S of minimum cardinality under the constraint that no block sum exceeds c. (In this problem, c is essentially a scale factor, so henceforth, without loss of generality, we make the usual normalization $c = 1$.)

Merely deciding whether a list of numbers can be partitioned into two blocks with equal sums is $\mathcal{NP}$-complete [Karp72], so all of the above problems are well-known to be $\mathcal{NP}$-complete. Thus, it is generally believed that algorithms solving these problems both exactly and efficiently do not exist. (Garey and Johnson [GJ79] provide a comprehensive treatment of $\mathcal{NP}$-completeness and its implications.) A substantial literature has built up over the past 20 years on the design and analysis of heuristic or approximation algorithms.

Most of the early research in the analysis of sequencing and packing algorithms concentrated on combinatorial, worst-case results. For example, if $H(S,m)$ denotes the makespan under heuristic H with S and m as inputs, the objective was to find for each m a least upper bound on $H(S,m)/\text{OPT}(S,m)$ over all S, where OPT stands for an optimum algorithm, i.e., $\text{OPT}(S,m)$ denotes the solution value of the makespan problem for the problem instance (S,m). Similar performance ratios were defined and studied in this way for bin-packing problems.

By contrast, the probabilistic analysis of algorithms covered in this book is a more recent line of research. In this type of analysis the elements of S are usually assumed to be n independent, identically distributed (i.i.d.) random variables with a given distribution function $F(x)$. The structure of most heuristics makes it convenient to suppose that these random variables are provided in the form of a list $L_n = (X_1, \ldots, X_n)$. Thus in the following a list L_n denotes an instance of bin packing with $c = 1$ assumed, and (L_n, m) denotes an instance of the makespan scheduling problem. For a given algorithm H, $H(L_n, m)$ and $H(L_n)$ are the obvious random variables, whose distributions are the goals of a probabilistic analysis.

Often, however, complete knowledge of distributions seems to be out of reach, and we must settle for weaker results, such as asymptotic statements about expected values and perhaps higher moments. Results are expressed as functions of n with $c = 1$ assumed for bin packing, and both n and m for the partition problem. Since the performance of an approximation algorithm should be assessed relative to the best possible performance, namely, that of an optimization algorithm, a great deal of the analysis in this book goes into the study of optimum performance.

As might be expected, a probabilistic analysis is frequently quite difficult, even when one is only attempting to find expected values. A principal source of difficulty is created by the conditional probabilities that arise in the analysis of an algorithm. For example, if an algorithm begins by comparing X_1 and X_2, then in the branch of the algorithm corresponding to $X_1 > X_2$ the

1.1. OVERVIEW

distributions are conditioned upon the event $\{X_1 > X_2\}$. After the algorithm has made a substantial number of comparisons, the conditioning can easily make a direct analysis formidable. Thus a primary goal of this book is to describe the various methods that have been used to deal with difficulties of this kind.

The focus on methodology limits our coverage of previous research to those results that we feel are good illustrations of the analytical tools that have been used, especially those that promise wide applicability in solving related problems not covered in this book. The literature is quite extensive, and even within this theme we cannot claim to make a complete survey of the available tools. More thorough reviews of results on the large variety of problems related to packing and partitioning can be found in several surveys. Dyckhoff [Dyck90] references these surveys in a recent article that also proposes a general taxonomy of such problems. The survey by Coffman, Garey, and Johnson [CGJ84] and the annotated bibliography by Karp et al. [KLMR84] are particularly appropriate sources of extensions and variants of the problems/algorithms studied in this book. We also remark that there is a sizeable literature, which we do not cover, on heuristic methods more elaborate than many of those covered here, e.g., local-neighborhood search, simulated annealing, and branch-and-bound methods. As we will see, even the analysis of relatively simple heuristics can often pose a substantial challenge.

The remainder of this chapter discusses examples of application areas for the abstract problems of packing and partitioning, describes several classical approximation algorithms that will receive special emphasis in later chapters, and discusses some notation to be used throughout the book.

Chapter 2 is the foundation for the remaining chapters; it introduces and illustrates a number of basic problem-solving techniques. Some of these are quite general and widely used; others are more specialized and hence more limited in their application. A number of fundamental, often-used background results in applied probability are also presented. Chapter 3 continues the discussion of background results by discussing matching problems in the plane, which have found numerous applications in the analysis of packing and other problems.

Chapter 4 applies the techniques of Chapter 2 to the analysis of algorithms for the scheduling (makespan) and partitioning problems. Chapters 5 and 6 deal similarly with the analysis of one-dimensional bin packing, the first of these focusing on the optimum solution and the second on approximation algorithms. Chapter 7 concludes the book with a discussion of packing problems in two dimensions.

1.2 Illustrative applications

The applications of packing and partitioning problems are huge in number and scope. We mention a few here to illustrate the variety, but the reader will have no difficulty in adding many more to the list.

Consider one-dimensional bin packing, for example. This problem arises in stock-cutting applications whenever material such as cable, lumber, pipe, or tape is supplied in a standard length; a list of demands for pieces of the material, each piece being no larger than the standard length, is to be satisfied by cutting up a minimum number of standard lengths, thereby minimizing the wasted material, also called the trim loss. By extending this notion of "material" to the more general concept of "resource," we find diverse applications in many different industries. For example, let the resource be time. We could be assigning commercials (pieces) to station breaks (fixed time slots) in the television or radio industry. Let the resource be weight. In the transportation industry we could be loading objects of varying weight onto a collection of trucks with identical weight limits.

In the design and programming of computers the resource could be storage. In table formatting, variable-length items of information are to be distributed among computer words, and in storage allocation records/files are to be assigned to equal-capacity disk cylinders.

The classical makespan problem originated in the scheduling requirements of industrial job shops and was given a fresh impetus by researchers concerned with parallel processing in computer systems. However, it bears repeating that one-dimensional bin packing and makespan scheduling are problems on the same basic structure. They differ only in which parameter is constrained (number of blocks or maximum block sum) and which parameter is taken as the objective function, so both can occur in the same applications. Makespan formulations tend to be design questions; for example, one can ask for the minimum truck design, in the sense of weight capacity, that is needed to transport a given collection of objects in a given number of trucks; the corresponding bin-packing problem is the minimum number of trucks of a prespecified weight capacity needed to transport the collection of objects.

The applications of packing in two dimensions are also ubiquitous. We begin with the study of a problem called *strip packing*. We are given a semi-infinite strip of unit width and a problem instance $L_n = (R_1, R_2, \ldots, R_n)$ specifying a list of rectangles $R_i = (X_i, Y_i)$, $1 \leq i \leq n$, with $X_i \leq 1$ the width and Y_i the height of R_i. The object is to pack the rectangles into the strip so that

1. the rectangles do not overlap each other or the edges of the strip,

2. the rectangles are packed with their sides parallel to the sides of the strip (90° rotations are disallowed), and

3. the packing height is minimized, where the *packing height* is the maximum height reached by the tops of the rectangles in a vertically oriented strip.

We also study a variant, called *two-dimensional bin packing*, where horizontal boundaries are placed at the integer heights of the strip. In this problem, strip packings are subject to the additional constraint that no rectangle can overlap any of these boundaries. In this version of strip packing, the "size" of the packing is taken to be the least integer no smaller than the height of the packing, i.e., the number of bins (unit squares) spanned by the packing.

Stock-cutting is again a far-reaching application area. The coils, spools, or lengths in one dimension become rolls, strips, or sheets of textiles, paper, plastics, sheet metal, etc., from which rectangular shapes are to be cut. Storage applications appear in countless settings; e.g., consider the archival storage of paintings (on walls) in large art museums, the allocation of advertisements in the yellow pages of a telephone directory, etc.

Scheduling under a resource constraint is another general applications area. For example, jobs (programs) in a computer have dimensions corresponding to both running time and storage requirements. The problem is to schedule a collection of jobs under a fixed limit on the total storage being occupied by jobs running in parallel. The constrained resource could be power, a situation that arises in scheduling experiments in spacecraft.

1.3 Notation

This section lays down a few notational conventions. As is common in the literature of mathematics and computer science, we let ln denote $\log_e$ and lg denote $\log_2$. In cases where the base does not matter (provided it is a constant greater than 1), we simply write log.

When writing expressions, we assume that the division operator "/" has lower precedence than multiplication; this enables us to write, for example, $1/pk$ instead of the more cumbersome $1/(pk)$.

We will often consider the uniform distribution, so it is convenient to let $U(a,b)$ denote the distribution uniform over $[a,b]$; the use of $U(0,1)$ will be especially frequent.

In the probabilistic analysis of packing and partitioning problems, one rarely finds an explicit formula for the quantity of interest. A variety of weaker but still informative statements can be made. Using bin packing as an example, suppose that the expected value $\mathsf{E}[H(L_n)]$ is being investigated. If we are unable to find an explicit formula for $\mathsf{E}[H(L_n)]$, we may still be able to find a function $f(n)$ such that $\mathsf{E}[H(L_n)] \sim f(n)$, i.e., the ratio $\mathsf{E}[H(L_n)]/f(n)$ tends to 1 as $n \to \infty$. Failing this, we may be able to find upper or lower bounds on the ratio, which can be conveniently described in terms of the O-, Ω-, and Θ-notation explained in [Knut76]. Let $g(n)$ and $f(n)$ be two functions of interest. If there are positive constants C and C' such that

$$Cf(n) \leq g(n) \leq C'f(n) \qquad (1.1)$$

with finitely many exceptions, then we write $g(n) = \Theta(f(n))$. (Note that Θ-notation enables us to disregard constant factors; thus, for example, the base of the logarithm in $\Theta(\log n)$ is immaterial as long as it is a constant greater than 1.) If even a statement of the form in (1.1) cannot be proved, we may be able to establish the left or right side of (1.1). If $|g(n)| \leq Cf(n)$ with finitely many exceptions, we write $g(n) = O(f(n))$; note that because we take the absolute value in the definition, O-notation can be applied to usefully describe a function that may assume negative values. Similarly, if $Cf(n) \leq g(n)$ with finitely many exceptions, then we write $g(n) = \Omega(f(n))$. One other notation that will occasionally be useful is $g(n) = o(f(n))$, which means that $\lim_{n\to\infty} g(n)/f(n) = 0$.

Occasionally we will make the statement that a *percentage* $\Omega(1)$ of the items in some set satisfy some property; in view of the above definition, this means that for all but finitely many n, the percentage of items in the set which satisfy the property is bounded below by some constant $c > 0$. (Note that a percentage is always trivially $O(1)$.)

The quantity that plays the role of $g(n)$ will often be the expected value of some random variable; however, we may not always be able to describe the expectation. For example, in some cases it may be possible to prove a strong bound on the median of some variable but not on its mean.

Sometimes we can make very strong statements. For example, instead of just showing that $\mathsf{E}[H(L_n)]/f(n) \to 1$ as $n \to \infty$, we may be able to characterize the asymptotic behavior of $\mathsf{E}[H(L_n)] - f(n)$. As a more concrete example, we may be able to show that the number of bins used to pack n

items is asymptotic to the total size of the items. It then becomes interesting to investigate the difference between $\mathsf{E}[H(L_n)]$ and the total size, i.e., the total amount of wasted space in the bins, to determine rates of convergence as a function of n.

1.4 Classical algorithms

In this section we record, for convenient future reference, several of the standard algorithms that have been applied to the partitioning and bin-packing problems. Hereafter, references to these problems will use terminology commonly found in the literature. The elements of L_n are called *items* and *tasks* in bin packing and makespan scheduling, respectively; blocks are called *bins* $B_1, B_2, \ldots$ and *processors*, $P_1, P_2 \ldots$, respectively; and the process of assigning elements to blocks is termed *packing* and *scheduling*, respectively.

Initial studies of the algorithms described here were confined to combinatorial, worst-case behavior. References to original articles are made where appropriate. A general treatment of the combinatorial theory of scheduling and packing algorithms can be found in the book [Coff76].

Algorithms can generally be divided into two major classes, on-line and off-line. To be called *on-line*, an algorithm must consider the items one at a time and make an irrevocable decision as to where each item is to be packed or scheduled before any data are available about subsequent items. In an *off-line* algorithm all item sizes are available for inspection before any decisions must be made about the packing. Often an off-line algorithm uses this information simply to sort the items into decreasing order of size, and then process them in that order. We further categorize on-line algorithms as open-end or closed-end. An *open-end* algorithm must pack items as they arrive without knowledge of the total number of items to be packed; a *closed-end* algorithm may make use of the total number of items to be packed when making decisions. (These distinctions will be especially important in Chapter 6.)

1.4.1 Makespan scheduling

The probabilistic analysis of scheduling started with the *list scheduling* (LS) rule [Grah66]. According to this simple rule, initially available processors $P_1, \ldots, P_m$ are assigned tasks in the order listed by the input $L_n = (X_1, X_2, \ldots, X_n)$. The ith task, X_i, is assigned to any processor whose finishing time is minimum in the schedule for the first $i - 1$ tasks. (Thus, in particular, the first $m' = \min(m,n)$ tasks may be assigned to the processors $P_1, \ldots, P_{m'}$ in any order.) An analysis of this rule will be presented in Section 2.2.

We will denote *order statistics* by parentheses around subscripts; thus $X_{(i)}$ will denote the ith smallest of the task times. A substantial improvement in the performance of the LS rule can be obtained by the simple expedient of processing the tasks in order of decreasing size. The LS rule applied to the ordered list $(X_{(n)}, X_{(n-1)}, \ldots, X_{(1)})$ is called the LPT rule, for *Largest Processing Time* first [Grah69]. Such a heuristic is an instance of the *greedy method*, in which we try to achieve or come close to a global optimum by making decisions that seem desirable in some local sense. Here, by scheduling large tasks first, we are hoping that the later, smaller, tasks will decrease any large differences between processor loads. Section 4.1 investigates the probabilistic behavior of this rule.

1.4.2 Bin packing

In this section we define several bin-packing algorithms. It is useful to classify bins into one of three categories: empty, active, and closed. An *empty* bin is one into which no items have been placed. After an item has been placed into a bin, we consider it *active* if it is still available for placement of additional items, and *closed* if it is not. (The notion of active bins will be especially important in Section 6.3.5.) In each of the bin-packing algorithms we discuss, we assume that there is an unbounded sequence of initially empty bins, $B_1, B_2, \ldots$.

We first describe three on-line algorithms. In each of these, items of L_n are packed in turn, starting with X_1. The simplest is the linear-time algorithm *Next Fit*, abbreviated NF. (By a *linear-time algorithm* we mean one which uses $O(n)$ time.) Suppose $X_1, \ldots, X_{i-1}$, $i \geq 1$, have already been packed, and B_j is the highest indexed nonempty bin. Then X_i is placed in B_j if it fits, i.e., if X_i plus the sum of the sizes of items already in B_j is at most 1; otherwise, we close B_j, and X_i is placed into B_{j+1}. Note that this algorithm never has more than one active bin at a time. A study of the worst-case behavior of NF and certain of its variants can be found in [John74].

Two other rules, *First Fit* (FF) and *Best Fit* (BF), never close a bin as long as items remain to be packed. In fact, neither will start a new bin unless the next item to be packed will not fit into any of the bins used so far; where FF and BF differ is in the rule used to select among the active bins when more than one of them can accommodate the item. Specifically,

1.4. CLASSICAL ALGORITHMS

FF packs each item into the lowest indexed bin into which it fits, whereas BF packs X_i into a bin which can accommodate it with the smallest capacity left over (with ties resolved in favor of the lower-indexed bins). Although a naïve implementation of either of these would use $\Theta(n^2)$ worst-case time, by selection of an appropriate data structure [Knut73, Stan80] each can be implemented to run in $\Theta(n \log n)$ worst-case time. A worst-case analysis of FF and BF packings appears in [JDUGG74].

Just as in Section 1.4.1, we can try to improve the performance of NF, FF, or BF by first sorting the items to be packed into order of decreasing size, i.e., processing them in the order $X_{(n)}, X_{(n-1)}, \ldots, X_{(1)}$. The word *Decreasing* after the name of the heuristic (e.g., First Fit Decreasing) or the letter D after the acronym (e.g., FFD) indicates that this sorting is to be done. Note that each of NFD, FFD, and BFD is off-line since the sorting process examines each item before any placement is done. Worst-case results appear in [BC81] for NFD and in [JDUGG74] for FFD and BFD.

Chapter 2

Analysis Techniques

This chapter describes a number of basic approaches to the analysis of packing and partitioning problems and presents concrete illustrations for most of the methods. Additional, more elaborate instances of most of the approaches will be found in later chapters.

The background in applied probability needed to follow the methods introduced here is modest. However, it is worth noting that heavy reliance is placed on classical inequalities (which is hardly surprising in a book where bounds on algorithm performance are so often the objective). Boole's inequality is used so frequently that we mention it here explicitly: if $\{E_i\}$ is an arbitrary countable sequence of events, then $\Pr\{\bigcup_i E_i\} \leq \sum_i \Pr\{E_i\}$. Examples of other relevant combinatorial or probabilistic inequalities include the geometric-arithmetic mean inequality, Chebyshev's inequality, Schwarz' (or the Cauchy-Schwarz) inequality, Jensen's inequality, etc. The reader is assumed to be familiar with these results, at least in their simpler forms. (A listing of some basic combinatorial inequalities can be found in [AS70, Section 3.2], and probabilistic results can be found in [Fell68, Fell71]; the appendices of [Hofr87] also provide a useful list of such results.)

In this and the remaining chapters, particularly frequent use is made of probability bounds and limit laws for sums of i.i.d. random variables, and for the special cases corresponding to simple random walks, where random variables take the values ± 1 only. For this reason, the present chapter begins with a special section devoted to these results. The theorems of this section are not intended as an exhaustive survey of the available theory, but rather should be taken simply as lemmas for the results to follow in later sections and chapters.

As a final comment before getting into the analysis, we note the valuable role that numerical experimentation can play in guiding the probabilistic analysis of algorithms. In order to get a feeling for the probabilistic behavior of a problem, one can generate problem instances at random, and it may be sufficient to examine only a relatively small set of such instances. Thus it may well be possible to generate data for problems in which there are many thousands of items, and develop intuition about the asymptotic behavior of the solution. This can then guide the analysis. For examples of such experiments, the reader is referred to the work of Ong, Magazine, and Wee [OMW84], Bentley et al. [BJLM83], and McGeoch [McGe87].

2.1 Sums of i.i.d. random variables

Reducing the analysis of an algorithm to that of a sum of i.i.d. random variables often presents a difficult problem. But when this can be done, a large literature of useful results can be exploited. This section presents a sampling of these results that will be used on many occasions. It also illustrates one type of analysis commonly used to prove probability bounds for sums of bounded random variables.

2.1.1 Small deviations and the central limit theorem

Let $\mathfrak{n}(x)$ denote the density of the normal distribution with zero mean and unit variance, i.e.,

$$\mathfrak{n}(x) = \frac{1}{\sqrt{2\pi}} e^{-x^2/2},$$

and let $\mathfrak{N}(x)$ denote the corresponding distribution function

$$\mathfrak{N}(x) = \int_{-\infty}^{x} \mathfrak{n}(x)\, dx.$$

We note that estimates of the tails $1 - \mathfrak{N}(x)$ are available:

$$1 - \mathfrak{N}(x) \leq \mathfrak{n}(x)/x \quad \text{for } x > 0; \tag{2.1}$$

more precise asymptotic descriptions can be found in [AS70, Section 26.2].

Now let $X_1, X_2, \ldots$ be independent samples of a random variable X with distribution function $F(x)$, mean μ, and positive variance σ^2. For $n \geq 1$, call $S_n = \sum_{i=1}^{n} X_i$ the nth *partial sum* of X, and call

$$\hat{S}_n = S_n - n\mu,$$

2.1. SUMS OF I.I.D. RANDOM VARIABLES

which has zero mean, the nth *centered partial sum* of X. Call

$$\tilde{S}_n = \frac{S_n - n\mu}{\sqrt{n}\sigma},$$

with zero mean and unit variance, the nth *normalized partial sum* of X. Under appropriate conditions on $F(x)$, the central limit theorem states that the random variables $\tilde{S}_n$ converge in law to a normal random variable with zero mean and unit variance, i.e.,

$$\Pr\left\{\tilde{S}_n \leq x\right\} \to \mathfrak{N}(x) \quad \text{as} \quad n \to \infty. \tag{2.2}$$

A condition sufficient for (2.2) to hold is that the mean and variance of X exist and are finite [Fell71, Section VIII.4, Theorem 1]. Indeed, in many of our applications, X satisfies a much stronger condition: it has bounded range.

Estimates of the rate of convergence in (2.2) are also of interest. The following, which we state without proof, will sometimes serve our needs.

Theorem 2.1 (Berry-Esséen [Fell71, Section XVI.5]) *If μ, σ, and $\rho = \mathsf{E}[|X - \mu|^3]$ are bounded, then*

$$\left|\Pr\{\tilde{S}_n \leq x\} - \mathfrak{N}(x)\right| \leq \frac{3\rho}{\sigma^3 \sqrt{n}}. \tag{2.3}$$

Note that this theorem holds in a very general context, but provides no information about the density of the partial sums; in fact, the conditions do not imply that such a density exists. With somewhat stronger conditions the density can be estimated. Let $\varphi(\zeta) = \mathsf{E}[e^{i\zeta X}]$ be the *characteristic function* of X, and let $\mu_k = \mathsf{E}[(X - \mu)^k]$ be the kth *central moment* of X.

Theorem 2.2 (see [Fell71, Section XVI.2]) *Let X be a random variable with characteristic function φ. If μ, σ^2, μ_3, and μ_4 are bounded and there is a $\nu \geq 1$ such that $|\varphi|^\nu$ is integrable, then for $n \geq \nu$ the centered partial sum $\hat{S}_n$ has a density f_n satisfying*[1]

$$f_n(x) = \frac{1}{\sigma\sqrt{n}}\left(1 + \frac{1}{\sqrt{n}}P\left(\frac{x}{\sigma\sqrt{n}}\right)\right)\mathfrak{n}\left(\frac{x}{\sigma\sqrt{n}}\right) + O(n^{-3/2}),$$

where

$$P(z) = \frac{\mu_3}{6\sigma^3}(z^3 - 3z).$$

The hidden constants in the O-notation are independent of x.

---
[1] This is in a somewhat different form from that appearing in [Fell71]. The function f_n is scaled differently, and we have used the fact that for any fixed polynomial p the function $p(x)\mathfrak{n}(x)$ is uniformly bounded.

In fact, an asymptotic expansion for f_n, sometimes called the *Edgeworth expansion*, is available; see [Fell71, Section XVI.2].

Occasionally, it is convenient to be able to estimate the value $\Pr\{X = k\}$, where X has a binomial distribution. Theorem 2.1 is not suitable for this task since it only gives bounds on the partial sums $\Pr\{X \leq k\} = \sum_{i=0}^{k} \Pr\{X = i\}$, and Theorem 2.2 is not suitable since its conditions are not met. Fortunately, a convenient bound is available.

Theorem 2.3 ([Fell68, Section VII.2–VII.3]) *Let X be binomially distributed, giving the number of successes in n Bernoulli trials with each having success probability p; let $q = 1 - p$. Then*

$$\Pr\{X = k\} = \big(1 + o(1)\big) \frac{1}{\sqrt{npq}} \mathfrak{n}\left(\frac{k - np}{\sqrt{npq}}\right), \tag{2.4}$$

uniformly in k, provided k varies with n in such a way that $k - np = o(n^{2/3})$.

Note that virtually all of the mass of the distribution in Theorem 2.3 lies within the region where the estimate of (2.4) applies. In particular, suppose we choose an $\alpha \in (1/2, 2/3)$, say, $\alpha = 3/5$. Then since $\Pr\{X = k\}$ decreases as we move away from the center of the distribution, we have

$$\Pr\{|X - np| \geq n^\alpha\} = O(n) \Pr\{|X - np| = \lfloor n^\alpha \rfloor\}$$
$$= O(\sqrt{n}) \mathfrak{n}\big(\Theta(n^{\alpha - 1/2})\big)$$
$$= o(n^{-l}) \tag{2.5}$$

for any positive integer l. We will say such a quantity goes to zero quickly enough to *swallow polynomials*.

By letting $p = \frac{1}{2}$ in Theorem 2.3, we obtain an estimate of the binomial coefficients:

$$2^{-n} \binom{n}{k} = \big(1 + o(1)\big) \frac{1}{\sqrt{n/4}} \mathfrak{n}\left(\frac{k - n/2}{\sqrt{n/4}}\right) = \big(1 + o(1)\big) \sqrt{\frac{2}{n\pi}} e^{-2(k - n/2)^2/n}, \tag{2.6}$$

provided $k - n/2 = o(n^{2/3})$.

When applying these estimates, it is often useful to be able to approximate a sum by an integral. A powerful tool for this purpose is Euler's summation formula. For our purposes the following very special case will be sufficient. Assuming that f is differentiable over $[a, b]$, we have

$$\sum_{k=a}^{b} f(k) = \int_{a}^{b} f(x)\, dx + \frac{1}{2}\big(f(a) + f(b)\big) + e_n, \tag{2.7}$$

2.1. SUMS OF I.I.D. RANDOM VARIABLES

where
$$|e_n| \leq \frac{1}{2}\int_a^b |f'(x)|\,dx.$$
For a much more general form see [Knut73, Section 1.2.11.2]. Note that if a, b, and f vary as functions of n, but the number of sign changes in $f'(x)$ over $[a, b]$ is bounded by some absolute constant, we can conclude

$$\sum_{k=a}^{b} f(k) = \int_a^b f(x)\,dx + O\left(\max_{a \leq x \leq b} |f(x)|\right). \tag{2.8}$$

This observation will often be useful in our analyses.

2.1.2 Bounds on the tails of the distributions

Results of the type shown in Theorems 2.1 and 2.2 of the previous section are helpful when we are interested in estimating the probability of small deviations from the mean. When we are interested in showing that the probability of a large deviation is extremely small, these theorems are generally not useful, because of the large error bounds. We turn next to Chernoff estimates for the tails $\Pr\{\hat{S}_n > x\}$ for large x. Many such bounds begin with the following simple observation. For any event A, let $\mathbf{1}_A$ be the *indicator* for A, that is, the random variable that is 1 when A holds and 0 when it does not. One easily sees that for an arbitrary random variable Z,

$$\mathbf{1}_{\{Z \geq a\}} \leq e^{\lambda(Z-a)} \tag{2.9}$$

for any $\lambda \geq 0$.

Now let X be a random variable with finite expectation $\mu = \mathsf{E}[X]$, and let S_n be the sum $X_1 + X_2 + \cdots + X_n$ of n independent random variables distributed as X. Let a be a value greater than or equal to μ. Letting S_n play the role of Z in (2.9), and taking expectations, we obtain

$$\Pr\{S_n \geq na\} = \mathsf{E}[\mathbf{1}_{\{S_n \geq na\}}] \leq \mathsf{E}\left[e^{\lambda(S_n-na)}\right] = e^{-na\lambda}\mathsf{E}\left[\prod_{i=1}^n e^{\lambda X_i}\right]. \tag{2.10}$$

Now because of the independence of the X_i, we can rewrite this as

$$\Pr\{S_n \geq na\} \leq e^{-na\lambda}\prod_{i=1}^n \mathsf{E}\left[e^{\lambda X_i}\right] = \left(\mathsf{E}[e^{\lambda(X-a)}]\right)^n. \tag{2.11}$$

Recalling that this holds for any $\lambda \geq 0$, and letting $u(a) = \inf_{\lambda \geq 0} \mathsf{E}[e^{\lambda(X-a)}]$, we have

$$\Pr\{S_n \geq na\} \leq \bigl(u(a)\bigr)^n. \tag{2.12}$$

In fact, one can drop the constraint $\lambda \geq 0$ on the range of λ in the definition of $u(a)$ since by the inequalities $e^x \geq 1 + x$ and $a \geq \mu$ we would have

$$\lambda < 0 \implies \mathsf{E}[e^{\lambda(X-a)}] \geq \mathsf{E}[1 + \lambda(X-a)] = 1 + \lambda(\mu - a) \geq 1 = \mathsf{E}[e^{0(X-a)}].$$

A symmetric argument holds for the case $a < \mu$, so we have the following:

Theorem 2.4 ([Cher52]) *Suppose $\mu = \mathsf{E}[X]$ exists and is finite, and let S_n be distributed as the sum of n independent samples of X. Then defining*

$$u(a) = \inf_{\lambda} \mathsf{E}[e^{\lambda(X-a)}],$$

we have

$$a \geq \mu \implies \Pr\{S_n \geq na\} \leq \bigl(u(a)\bigr)^n,$$

and

$$a \leq \mu \implies \Pr\{S_n \leq na\} \leq \bigl(u(a)\bigr)^n.$$

Chernoff [Cher52] also shows that $u(a)$ is the best possible base in an exponential bound of this sort, in the sense that for $0 < b < u(a)$,

$$b^n = o(\Pr\{S_n \geq na\}) \quad \text{and} \quad b^n = o(\Pr\{S_n \leq na\}).$$

As an example of Theorem 2.4, suppose X has an exponential distribution with mean 1, i.e., $\Pr\{X \leq x\} = 1 - e^{-x}$ for $x \geq 0$. Then one readily computes

$$\mathsf{E}[e^{\lambda(X-a)}] = \int_0^\infty e^{\lambda(x-a)} e^{-x}\, dx = \frac{e^{-a\lambda}}{1-\lambda} \quad \text{for } \lambda < 1$$

(with divergence to $+\infty$ for greater λ), so for this distribution we have

$$u(a) = \inf_{\lambda < 1} \frac{e^{-a\lambda}}{1-\lambda} = ae^{1-a}, \tag{2.13}$$

where the minimum was determined to occur at $\lambda = 1 - 1/a$ by simple calculus. The resulting bound will be used in Section 4.1.

A similar sort of bound can be obtained with an assumption weaker than independence of the X_i [Hoef63]. A sequence $\hat{S}_i$, $i = 1, \ldots, n$, is said to be a *martingale* if

$$\mathsf{E}[\hat{S}_i \mid \hat{S}_1, \hat{S}_2, \ldots, \hat{S}_j] = \hat{S}_j, \quad \text{for } 1 \leq j \leq i \leq n.$$

Suppose we do not know the distributions of the X_i and are not even given that they are i.i.d. Instead, we know that their range is limited by

$$|X_i| \leq 1 \tag{2.14}$$

2.1. SUMS OF I.I.D. RANDOM VARIABLES

and that the sequence of centered partial sums $\hat{S}_i$ is a martingale. In terms of the X_i, this latter condition tells us that although the X_i may not be independent,

$$\mathsf{E}[X_i \mid X_1, X_2, \ldots, X_{i-1}] = \mathsf{E}[X_i]; \tag{2.15}$$

in other words, conditioning X_i on the values of any of the X_j, $j < i$, may affect its distribution but not its mean. Again, we wish to bound the probability that S_n is substantially greater than its mean. We can no longer use independence as we did to go from (2.10) to (2.11), since without it we cannot conclude for a function g that we have

$$\mathsf{E}\left[\prod_{i=1}^n g(X_i)\right] = \prod_{i=1}^n \mathsf{E}[g(X_i)]. \tag{2.16}$$

If, however, g is a linear function, (2.15) does imply (2.16). This motivates the following approach, which also takes advantage of (2.14): we bound each $e^{\lambda X_i}$ by an approximation linear in X_i:

$$e^{\lambda X_i} \leq \tfrac{1}{2} e^{-\lambda}(1 - X_i) + \tfrac{1}{2} e^{\lambda}(1 + X_i) \quad \text{for } |X_i| \leq 1.$$

(See Figure 2.1.) Now letting $\gamma_i = \mathsf{E}[X_i]$, we may write

$$\Pr\{\hat{S}_n \geq t\} \leq \mathsf{E}\left[e^{\lambda(\hat{S}_n - t)}\right] \leq e^{-\lambda t} \mathsf{E}\left[e^{\sum_{i=1}^n \lambda(X_i - \gamma_i)}\right]$$

$$= e^{-\lambda t} \mathsf{E}\left[\prod_{i=1}^n e^{\lambda X_i - \gamma_i}\right]$$

$$\leq e^{-\lambda t} \mathsf{E}\left[\prod_{i=1}^n e^{-\lambda \gamma_i}\left(\tfrac{1}{2} e^{-\lambda}(1 - X_i) + \tfrac{1}{2} e^{\lambda}(1 + X_i)\right)\right]$$

$$= e^{-\lambda t} \prod_{i=1}^n \mathsf{E}\left[e^{-\lambda \gamma_i}\left(\tfrac{1}{2} e^{-\lambda}(1 - X_i) + \tfrac{1}{2} e^{\lambda}(1 + X_i)\right)\right]$$

$$= e^{-\lambda t} \prod_{i=1}^n e^{-\lambda \gamma_i}\left(\tfrac{1}{2} e^{-\lambda}(1 - \gamma_i) + \tfrac{1}{2} e^{\lambda}(1 + \gamma_i)\right). \tag{2.17}$$

We will be finished if we can find a good bound on the multiplicands in the above product. To this end, let

$$\phi(\lambda, \mu) = e^{-\lambda \mu}\left(\tfrac{1}{2} e^{-\lambda}(1 - \mu) + \tfrac{1}{2} e^{\lambda}(1 + \mu)\right),$$

and let

$$L(\lambda, \mu) = \ln \phi(\lambda, \mu) = -(1 + \mu)\lambda + \ln\left(\frac{1 - \mu}{2} + \frac{1 + \mu}{2} e^{2\lambda}\right).$$

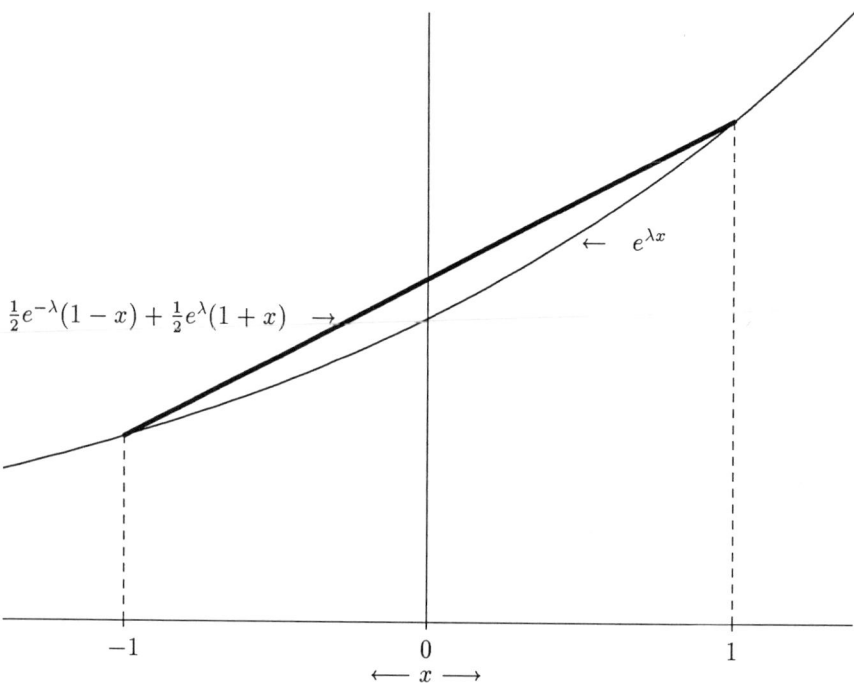

Figure 2.1: Illustration of the argument used in the proof of Hoeffding's bound. Since we know x ranges only over $[-1, 1]$, and since the exponential function is convex, we can bound $e^{\lambda x}$ by the bold straight line.

2.1. SUMS OF I.I.D. RANDOM VARIABLES

To bound this, note that

$$\frac{\partial}{\partial \lambda}L(\lambda,\mu) = -(1+\mu) + \frac{1+\mu}{\frac{1-\mu}{2}e^{-2\lambda} + \frac{1+\mu}{2}},$$

so $\frac{\partial}{\partial \lambda}L(0,\mu) = 0$, and

$$\frac{\partial^2}{\partial \lambda^2}L(\lambda,\mu) = \frac{(1+\mu)(1-\mu)e^{-2\lambda}}{\left(\frac{1+\mu}{2} + \frac{1-\mu}{2}e^{-2\lambda}\right)^2} \leq 1,$$

where we have used the geometric-arithmetic mean inequality. Hence by Taylor's theorem we have $L(\lambda,\mu) \leq \lambda^2/2$ so $\phi(\lambda,\mu) \leq e^{\lambda^2/2}$. Thus we can simplify (2.17) to

$$\Pr\{\hat{S}_n \geq t\} \leq e^{-\lambda t}\prod_{i=1}^{n} e^{\lambda^2/2} = e^{-\lambda t + n\lambda^2/2}.$$

It is routine to show that the right side is minimized by the choice $\lambda = t/n$, which leads to the bound $e^{-t^2/2n}$. Letting $t = x\sqrt{n}$, we obtain (2.18) in the following theorem. Replacing X_i by $-X_i$ gives (2.19).

Theorem 2.5 ([Hoef63]) *Let X_i, $i = 1, 2, \ldots, n$, be independent random variables that assume values in the range $|X_i| \leq 1$. Let $S_n = X_1 + X_2 + \cdots + X_n$, and $\hat{S}_n = S_n - \mathsf{E}[S_n]$. Then for $x \geq 0$*

$$\Pr\{\hat{S}_n \geq x\sqrt{n}\} \leq e^{-x^2/2}, \tag{2.18}$$

and

$$\Pr\{\hat{S}_n \leq -x\sqrt{n}\} \leq e^{-x^2/2}. \tag{2.19}$$

This conclusion remains valid even if we weaken the assumption that the X_i are independent to the assumption that the sequence $\hat{S}_n$ is a martingale.

In some cases we require estimates in which the dependence on parameters of the distribution of X is explicit. A convenient such estimate, known as Bernstein's bound [Hoef63, Equation (2.13)], is given below.

Theorem 2.6 (Bernstein) *Let X_i, $i = 1, 2, \ldots, n$, be independent random variables with variance σ^2, and suppose that $|X_i - \mathsf{E}[X_i]| \leq M$. Then for $x \geq 0$*

$$\Pr\{\hat{S}_n \geq x\sqrt{n}\} \leq \exp\left\{-\frac{x^2/2}{\sigma^2 + \frac{Mx}{3\sqrt{n}}}\right\}.$$

Sometimes we want to be able to bound the probability that *any* of the partial sums $\hat{S}_i$ exceeds some limit. This can, for example, be very useful when a random walk arises. A first attempt to obtain such a bound might simply use Theorem 2.5 and Boole's inequality to obtain

$$\Pr\left\{\max_{1\le i\le n} \hat{S}_i \ge t\right\} = \Pr\left\{\bigcup_{i=1}^{n} \{\hat{S}_i \ge t\}\right\}$$

$$\le \min\left\{1, \sum_{i=1}^{n} \Pr\{\hat{S}_i \ge t\}\right\}$$

$$\le \min\left\{1, \sum_{i=1}^{n} e^{-t^2/2i}\right\}.$$

This bound turns out to be rather weak, however; in fact, some computation shows that the value of t for which it first drops below 1 is $\Theta(\sqrt{n \log n})$. By a more careful argument combining Skorohod's inequality (see [Brei68, Section 3.4, Lemma 3.21]) with Theorem 2.5 we can obtain a stronger result:

Theorem 2.7 *Let X_i, $i = 1, 2, \ldots, n$, be independent random variables that assume values in the range $|X_i| \le 1$. Let $S_i = X_1 + X_2 + \cdots + X_i$, and $\hat{S}_i = S_i - \mathsf{E}[S_i]$. Then for $x \ge 0$*

$$\Pr\left\{\max_{1\le i\le n} \hat{S}_i \ge x\sqrt{n}\right\} \le 2e^{-x^2/8}, \text{ and}$$
$$\Pr\left\{\min_{1\le i\le n} \hat{S}_i \le -x\sqrt{n}\right\} \le 2e^{-x^2/8}. \qquad (2.20)$$

Proof. Let $2t = x\sqrt{n}$, and let C_j be the event that $\hat{S}_j$ is the first centered partial sum greater than or equal to $2t$, i.e.,

$$C_j = \{\hat{S}_j \ge 2t \text{ and } \hat{S}_k < 2t \text{ for all } 1 \le k < j\}.$$

Then

$$\left\{\max_{1\le i\le n} \hat{S}_i \ge 2t\right\} = \bigcup_{j=1}^{n} C_j,$$

and[2]

$$\Pr\left\{\max_{1\le i\le n} \hat{S}_i \ge 2t\right\} \le \sum_{j=1}^{n} \Pr\{C_j\}. \qquad (2.21)$$

---

[2]In fact, since the C_j are disjoint, equality holds in (2.21), but we do not need equality for this proof.

2.1. SUMS OF I.I.D. RANDOM VARIABLES

Using the disjointness of the C_j and the fact that

$$\{C_j\} \cap \{\hat{S}_n - \hat{S}_j \geq -t\} \subseteq \{\hat{S}_n \geq t\},$$

we can write

$$\Pr\{\hat{S}_n \geq t\} \geq \sum_{j=1}^{n} \Pr\{C_j \text{ and } \hat{S}_n - \hat{S}_j \geq -t\}$$

$$= \sum_{j=1}^{n} \Pr\{C_j\} \Pr\{\hat{S}_n - \hat{S}_j \geq -t\}, \qquad (2.22)$$

where in the last step we have used the independence of the events $\{C_j\}$ and $\{\hat{S}_n - \hat{S}_j \geq -t\}$. Now by Theorem 2.5

$$\Pr\{\hat{S}_n \geq t\} \leq e^{-t^2/2n}$$

and

$$\Pr\{\hat{S}_n - \hat{S}_j \geq -t\} = 1 - \Pr\{\hat{S}_n - \hat{S}_j < -t\} \geq 1 - e^{-t^2/2(n-j)} \geq 1 - e^{-t^2/2n}.$$

Substituting these bounds into (2.22), we obtain

$$e^{-t^2/2n} \geq \left(1 - e^{-t^2/2n}\right) \sum_{j=1}^{n} \Pr\{C_j\}. \qquad (2.23)$$

From (2.21) and (2.23) we get

$$\Pr\left\{\max_{1 \leq i \leq n} \hat{S}_i \geq x\sqrt{n}\right\} \leq \min\left\{1, \frac{e^{-x^2/8}}{1 - e^{-x^2/8}}\right\} \leq 2e^{-x^2/8},$$

which is the first part of (2.20). The second part follows upon replacing X_i by $-X_i$. ∎

Note that an integration of the above tail probability yields

$$\mathsf{E}\left[\max_{1 \leq i \leq n} \hat{S}_i\right] = O(\sqrt{n}).$$

In all cases of interest here, we will have a matching lower bound of $\Omega(\sqrt{n})$. A demonstration of this fact for a specific distribution appears in the next section.

2.1.3 Estimates of moments

In addition to the above convergence in distribution, it is also helpful to know whether the moments converge to the corresponding moments of the limit distribution. Here, we often deal with i.i.d. bounded random variables X_i, $i = 1, 2, \ldots$; for this case the desired convergence is easy to prove. The proof gives us a technical application of the Hoeffding bounds in Theorem 2.5.

Let S have the normal distribution with zero mean and unit variance, so that, as $n \to \infty$, $\tilde{S}_n$ converges in distribution to S. The sequence $|\tilde{S}_n|^r$, $n \geq 1$, is said to be *uniformly integrable* if, uniformly in n,

$$\lim_{c \to \infty} \mathsf{E}[|\tilde{S}_n|^r \cdot 1_{|\tilde{S}_n| \geq c}] = 0. \tag{2.24}$$

If this holds, then we have (see [Serf80, p. 13])

$$\mathsf{E}[|\tilde{S}_n|^r] \to \mathsf{E}[|S|^r] \quad \text{and} \quad \mathsf{E}[\tilde{S}_n^r] \to \mathsf{E}[S^r].$$

To show that (2.24) holds, consider first the integral

$$\int_c^\infty x^r \, d\Pr\{\tilde{S}_n \leq x\} = -\int_c^\infty x^r \, d\Pr\{\tilde{S}_n > x\}$$

$$= -x^r \Pr\{\tilde{S}_n > x\}\Big|_c^\infty + \int_c^\infty r x^{r-1} \Pr\{\tilde{S}_n > x\} \, dx. \tag{2.25}$$

Since the X_i are bounded, $\mathrm{Var}(\tilde{S}_n) = O(n)$, and by Theorem 2.5 there is a constant $\alpha > 0$ such that $\Pr\{\tilde{S}_n > x\} \leq e^{-\alpha x^2/2}$ uniformly in n. Using this bound in (2.25), we get easily

$$\lim_{c \to \infty} \int_c^\infty x^r \, d\Pr\{\tilde{S}_n \leq x\} = 0. \tag{2.26}$$

Now Theorem 2.5 also gives $\Pr\{\tilde{S}_n \leq -x\} \leq e^{-\alpha x^2/2}$, so by a similar argument

$$\lim_{c \to \infty} \int_{-\infty}^{-c} |x^r| \, d\Pr\{\tilde{S}_n \leq x\} = 0 \tag{2.27}$$

uniformly in n. Thus (2.24) follows at once from (2.26) and (2.27).

We illustrate these results with an example that also shows how to deal with certain situations where the number of random variables being summed is also random. Let X_i, $i = 1, 2, \ldots$, be i.i.d. random variables with $\Pr\{X_i = +1\} = \Pr\{X_i = -1\} = p/2$ and $\Pr\{X_i = 0\} = 1 - p$ for some fixed $0 < p \leq 1$. Put $S_n = \sum_{i=1}^n X_i$ as usual, and observe that $\max(0, S_n)$ is equal in distribution to the excess of heads over tails after a sequence of N tosses of a fair coin, where N has a binomial distribution with parameters p and n (and hence mean pn). We have $\mathsf{E}[S_n] = 0$ and $\mathrm{Var}(S_n) = np$. Now

$$\max(0, S_n) = \frac{|S_n| + S_n}{2},$$

so by the uniform integrability of $|S_n|/\sqrt{np}$ we get

$$\mathsf{E}[\max(0, S_n)] \sim \frac{1}{\sqrt{2\pi np}} \int_0^\infty x e^{-x^2/2np}\, dx = \sqrt{np/2\pi} \qquad (2.28)$$

as $n \to \infty$. Also, since $|S_n| = \max(0, S_n) + \max(0, -S_n)$, we obtain by symmetry

$$\mathsf{E}[|S_n|] = 2\mathsf{E}[\max(0, S_n)] \sim \sqrt{2np/\pi} \qquad (2.29)$$

as $n \to \infty$.

The expectation of $\max_{1 \le i \le n} S_i$ is also of interest. Consider the case $p = 1$. An integration of the tail probability provided by Theorem 2.7 gives

$$\mathsf{E}\left[\max_{1 \le i \le n} \hat{S}_i\right] = O(\sqrt{n}).$$

Now, clearly, $\max_{1 \le i \le n} \hat{S}_i$ is bounded below by $\max(0, S_n)$. Hence by (2.28) we also have

$$\mathsf{E}\left[\max_{1 \le i \le n} \hat{S}_i\right] = \Omega(\sqrt{n}),$$

so we conclude

$$\mathsf{E}\left[\max_{1 \le i \le n} \hat{S}_i\right] = \Theta(\sqrt{n}). \qquad (2.30)$$

2.2 Markov chains

Techniques for obtaining partitions that approximately solve some problem have often been defined by sequential algorithms, such as those in Section 1.4. Thus a fundamental approach to the probabilistic analysis of such an algorithm begins with the formulation of an appropriate Markov chain representing the execution of the algorithm as it forms the partition element by element. A state of the Markov chain must represent block sums in a suitable way; given the state space, the transition function will be defined by the algorithm. If the state space and transition function are simple enough, then classical methods can be applied to obtain results for general n through a transient analysis or asymptotic results for large n through a steady-state analysis.

To illustrate ideas, let us consider an average-case analysis of the simple LS rule for the makespan problem, defined in Section 1.4.1. Let us consider the simplest nontrivial case, $m = 2$, and define V_i as the (positive) difference between the processor finishing times after the first i tasks have been scheduled. Thus

$$V_i = \begin{cases} |V_{i-1} - X_i| & \text{for } 1 \leq i \leq n, \\ 0 & \text{for } i = 0, \end{cases} \quad (2.31)$$

so the analysis of

$$\text{LS}(L_n, 2) = \frac{V_n}{2} + \frac{1}{2}\sum_{i=1}^{n} X_i \quad (2.32)$$

reduces to the analysis of the chain $\{V_i\}_{i \geq 0}$. It is easy to see that $\{V_i\}_{i \geq 0}$ is a Markov chain. Let F be the common distribution function of the X_i, and assume that $F(x)$ has a density $f(x)$ on $[0, \infty)$. Then the distributions of the V_i, $i \geq 1$, have densities $f_i(x)$ satisfying the recurrence

$$f_i(y) = \begin{cases} f(y) & \text{for } i = 1, \\ \int_0^\infty k(x, y) f_{i-1}(x)\, dx & \text{for } i \geq 2, \end{cases} \quad (2.33)$$

where $k(x, y)$ is the transition density of the Markov chain. A straightforward analysis of (2.31) shows that

$$k(x, y) = \begin{cases} f(x - y) + f(x + y) & \text{for } 0 < y < x, \\ f(x + y) & \text{for } y > x > 0. \end{cases} \quad (2.34)$$

From (2.33) and (2.34), a stationary density f_∞ in the limit $i \to \infty$ is defined by

$$f_\infty(y) = \int_0^\infty f_\infty(x + y) f(x)\, dx + \int_y^\infty f_\infty(x - y) f(x)\, dx$$

$$= \int_0^\infty f_\infty(x + y) f(x)\, dx + \int_0^\infty f_\infty(x) f(x + y)\, dx, \quad (2.35)$$

with a solution

$$f_\infty(y) = \frac{1 - F(y)}{\mathsf{E}[X]}. \quad (2.36)$$

As an example, suppose the distribution of X is $U(0, 1)$, so $F(x) = x$ over $[0, 1]$. Then (2.36) gives $f_\infty(y) = 2(1 - y)$ over $[0, 1]$. Equations (2.33) and (2.34) show that in fact $f_i(y) = 2(1 - y)$, $0 \leq x \leq 1$, for all $i \geq 2$. Hence (assuming $n \geq 2$) we have $\mathsf{E}[V_n] = 1/3$, so (2.32) gives

$\mathsf{E}[\mathrm{LS}(L_n, 2)] = (n/4)(1 + 2/3n)$. Then $\mathsf{E}[\mathrm{OPT}(L_n, 2)] \geq \mathsf{E}\left[\sum_{i=1}^{n} X_i\right]/2 = n/4$ yields the relative-performance bound

$$\frac{\mathsf{E}[\mathrm{LS}(L_n, 2)]}{\mathsf{E}[\mathrm{OPT}(L_n, 2)]} \leq 1 + \frac{2}{3n}.$$

Another application of the Markov-chain approach can be found in Section 6.3.1, where NF bin packing is analyzed. We point out that this approach has been successful only for the simpler, less efficient heuristics. The state spaces of Markov chains defined for heuristics like FF and BF appear to be too large and unwieldy for a tractable analysis.

2.3 Bounds

When approaches to exact results fail, one turns naturally to bounds in order to get at least some information. Of course, bounds can enter the analysis in many ways. Here, we consider bounds introduced at an early stage. A prime example consists of applying at the outset a bound to the objective function itself; e.g., in analyzing the bin-packing heuristic H it may be possible to find a useful, more tractable function $g(L_n)$ such that $g(L_n) \geq H(L_n)$ for all problem instances L_n. Thus, if $\mathsf{E}[g(L_n)]$ is relatively easy to calculate, then for the average-case performance measure we obtain the bound $\mathsf{E}[H(L_n)] \leq \mathsf{E}[g(L_n)]$.

A concrete example is provided by the LS algorithm, discussed in the previous section. Let the objective function be the *relative error*

$$R^{\mathrm{LS}}(L_n, m) = \frac{\mathrm{LS}(L_n, m) - \mathrm{OPT}(L_n, m)}{\mathrm{OPT}(L_n, m)}. \tag{2.37}$$

To develop a bound, observe first that a scheduling rule can do no better than to keep all the processors busy during the makespan, assigning to each processor $(1/m)$th of the total work. Thus

$$\mathrm{OPT}(L_n, m) \geq \frac{1}{m} \sum_{i=1}^{n} X_i. \tag{2.38}$$

On the other hand, consider any algorithm that does not introduce idle time when unfinished tasks remain; such an algorithm can do no worse than to finish off the schedule with the largest task executing alone, i.e., all but one processor is idle during the last $X_{(n)} = \max_{i=1}^{n} X_i$ time units. Then

$$\mathrm{LS}(L_n, m) \leq \frac{1}{m} \left((m-1) X_{(n)} + \sum_{i=1}^{n} X_i \right). \tag{2.39}$$

If we can assume $X_{(n)} \leq 1$, substitution of (2.38) and (2.39) into (2.37) yields the bound

$$R^{\text{LS}}(L_n, m) \leq \frac{m-1}{\sum_{i=1}^{n} X_i}, \qquad (2.40)$$

so our problem has been reduced to the analysis of sums of i.i.d. random variables.

To illustrate the possible results, assume the X_i are uniformly distributed on $[0, 1]$. Then an appeal to Theorem 2.5 with $x\sqrt{n} = n/4$, for example, yields

$$\Pr\left\{R^{\text{LS}}(L_n, m) > \frac{4(m-1)}{n}\right\} \leq \Pr\left\{\sum_{i=1}^{n}\left(X_i - \frac{1}{2}\right) < -\frac{n}{4}\right\} \leq e^{-n/32},$$

which gives information on how fast the relative error tends to 0 as $n \to \infty$.

The coefficient $1/32$ in the exponent can be increased substantially by exploiting the parameters of $U(0,1)$; e.g., see Theorem 2.6. Also, the distribution $U(0,1)$ allows a slight tightening of (2.40), since (2.38) and (2.39) imply $R^{\text{LS}}(L_n, m) \leq \frac{m-1}{1+Z_{n-1}}$, where $Z_{n-1} = \sum_{i=1}^{n-1} X_{(i)}/X_{(n)}$ is a sum of $n-1$ independent random variables from $U(0, 1)$. Coffman and Gilbert [CG85] give a detailed analysis of this bound.

The exponential case $F(x) = 1 - e^{-\lambda x}$, $x \geq 0$, was also studied in [CG85]. For this case, a more convenient bound was found to be $R^{\text{LS}}(L_n, m) \leq (m-1)V/(mW + V)$, where W is the earliest processor finishing time in the LS schedule and $V = \text{LS}(L_n, m) - W$ is the difference between the earliest and latest processor finishing times. This bound follows easily from $\text{OPT}(L_n) \geq (mW + V)/m$ and $\text{LS}(L_n, m) = W + V$. Using similar bounds, Boxma [Boxm85] extended the results to the case where $F(x)$ is the Erlang-k distribution.

2.4 Dominating algorithms

A common way to find a useful bound on $H(L_n)$ is to introduce a simpler, more easily analyzed algorithm H' for which it can be proved that $H'(L_n) \geq H(L_n)$ for each problem instance L_n—then we say that H' dominates H. Many of the results in the probabilistic analysis of bin packing afford good illustrations of this approach. Although the technique applies equally well to makespan scheduling, it has not been exploited there to nearly the same extent. As an illustration, we consider the Next Fit Decreasing (NFD) heuristic defined in Section 1.4.2.

Csirik et al. [CFFGR86] obtained results for NFD by analyzing a dominating algorithm called *sliced NFD with parameter r*, SNFD_r. SNFD_r is the same as NFD until it first packs an rth item into a bin. (Note that the size of such an item can be no greater than $1/r$.) At that point SNFD_r packs the remaining items r to a bin. For any list L_n, SNFD_r clearly dominates NFD, i.e., $\text{SNFD}_r(L_n) \geq \text{NFD}(L_n)$.

Let k_i, $i \geq 1$, denote the number of items in L_n with sizes in the interval $\bigl(1/(i+1), 1/i\bigr]$, and let $K_i = k_i + k_{i+1} + \cdots$. The following bounds are easily verified:

$$\sum_{i=1}^{r} \left\lfloor \frac{k_i}{i} \right\rfloor \leq \text{NFD}(L_n) \leq \text{SNFD}_r(L_n) \leq \left\lceil \frac{K_{r+1}}{r} \right\rceil + \sum_{i=1}^{r} \left\lceil \frac{k_i}{i} \right\rceil.$$

Now suppose item sizes are independent samples from $U(0,1)$. Then $\mathsf{E}[k_i] = n/i(i+1)$ and $\mathsf{E}[K_{r+1}] = n/(r+1)$. We now let $r \to \infty$, as a function of n, rapidly enough so that the contribution of K_{r+1}/r is $o(n)$, but slowly enough so that the number of times we round to an integer is also $o(n)$; for example, let $r = \sqrt{n}$. Then the above bounds show that

$$\mathsf{E}[\text{NFD}(L_n)] \sim \sum_{i=1}^{\infty} \frac{n}{i^2(i+1)} = \left(\frac{\pi^2}{6} - 1\right) n \quad \text{as } n \to \infty,$$

a result also obtained by [HK86]. [CFFGR86] also show that $\text{NFD}(L_n)$ satisfies a central limit theorem, that is, that in the limit of large n, $\text{NFD}(L_n)$ is normally distributed.

Finally, we mention results of Rhee [Rhee87], who showed that

$$\left| \text{NFD}(L_n) - \sum_{i=1}^{\infty} \frac{k_i}{i} \right| \leq 1 + \log n.$$

Now let $f(x) = 1/\lfloor 1/x \rfloor$, that is, $f(x)$ is x rounded up to the next value which is a reciprocal. Then $\sum_{i=1}^{\infty} k_i/i = \sum_{i=1}^{n} f(X_i)$. Using these observations Rhee easily achieves a central limit theorem for $\text{NFD}(L_n)$ for general distributions F on $[0,1]$.

2.5 Bounds that usually hold

In some cases it helps to employ a deterministic bound that may not hold in all cases. For example, instead of looking for a bound that always holds, it may be possible to find a better bound $g(n)$ such that $H(L_n) \leq g(n)$ holds in some very likely event A, and in the remaining cases a much weaker bound

$H(L_n) \leq g'(n)$ holds. If q_n is the probability of the complement $\bar{A}$ of A, and $q_n \to 0$ for large n sufficiently rapidly so that $q_n g'(n) = o(g(n))$, then we have

$$\mathsf{E}[H(L_n)] = \Pr\{A\}\mathsf{E}[H(L_n) \mid A] + \Pr\{\bar{A}\}\mathsf{E}[H(L_n) \mid \bar{A}]$$
$$\leq (1 - q_n)g(n) + q_n g'(n) \sim g(n). \tag{2.41}$$

As an example, we will give a simple proof that if the X_i are i.i.d. and uniform over some interval $[a,b] \subseteq [0,1]$ that is symmetric about $1/p$ for some integer $p > 1$, then $\mathsf{E}[\mathrm{OPT}(L_n)] \sim n/p$. (Other proofs of this result appear in [Luek83, RT88b]. The proof here combines ideas from both of these, and is similar to a proof involving triangular distributions appearing in [KLV87].) A lower bound of

$$\mathsf{E}[\mathrm{OPT}(L_n)] \geq n/p \tag{2.42}$$

is trivial since $\mathsf{E}[\sum_{i=1}^n X_i] = n/p$. For the upper bound we will describe a heuristic H and show, by the method of a bound that usually holds, that $\mathsf{E}[H(L_n)] \sim n/p$. It will be helpful to consider a further randomization of the data allowing us to view them as points distributed uniformly over a rectangle, as shown in Figure 2.2; the horizontal coordinate of a point is the item size, and the vertical components are chosen independently and uniformly. We divide the rectangle as follows. First, we divide it by a horizontal line so that there is a long narrow subrectangle at the top, with the ratio between the top and bottom part being $1 : p - 1$. Next, we divide the lower part horizontally into $p-1$ equal subrectangles. These subdivisions are shown as bold lines in the figure. Now, for some $k > 1$, we divide each of the subrectangles by vertical lines into k equal slices, giving us a total of pk slices. We partition the slices into $k+1$ classes, satisfying the condition that for each class, selecting one item from each slice in the class will always give us a set of items that fit together into a bin. To form class number j, $1 \leq j \leq k - 1$, we select the jth slice from the left in each of the lower $p - 1$ subrectangles, and the $(j+1)$st slice from the right in the top subrectangle. For $j = k$, class number j is formed similarly except that there is no representative in the top subrectangle. Finally, the top right slice goes into a special class $k + 1$ by itself. (See Figure 2.2.) It is not hard to verify that these classes satisfy the desired condition. Note that the total number of classes is $k+1$. Hence, if we let M be the maximum number of items appearing in any slice, an obvious heuristic, say, H, can pack all of the items using only $(k + 1)M$ bins.

Now fix k and some $\epsilon > 0$, and let A be the event that $M \leq (n/pk)(1+\epsilon)$. The bound that usually holds is the one we can conclude in the event that A occurs, i.e., in the notation of (2.41), $g(n) = (k+1)M \leq \frac{n}{p}\frac{k+1}{k}(1+\epsilon)$; when

2.5. BOUNDS THAT USUALLY HOLD

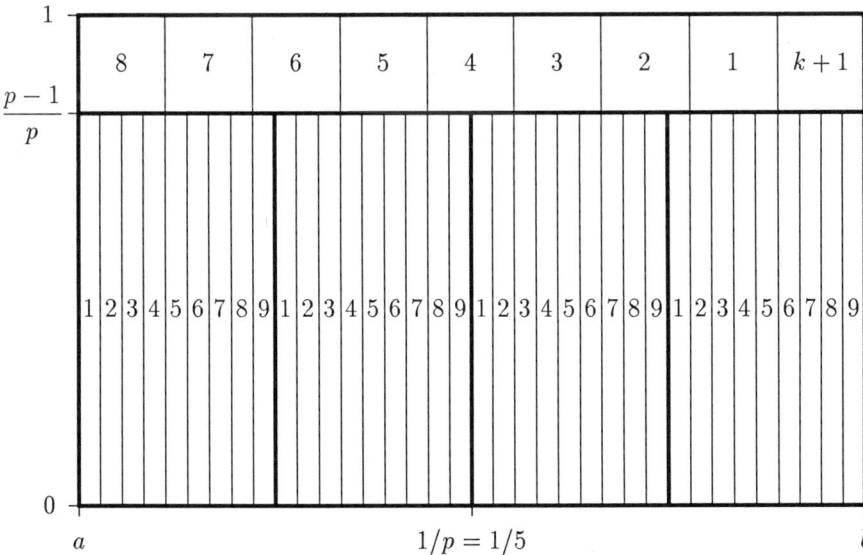

Figure 2.2: Packing about a reciprocal. This figure illustrates a simple proof that items distributed uniformly over an interval symmetric about a reciprocal $1/p$ can be packed using an average of n/p bins. Here $p = 5$, and the number k of slices per subrectangle is 9. The symbol appearing inside a slice indicates the class of which it is a member.

A fails, we fall back on the trivial bound $g'(n) = n$, which corresponds to using a separate bin for each item.

To estimate the probability q_n that A fails, we begin by considering a single slice and letting S_n be the number of items in that slice. Now the sizes of the items are independent, and each falls in the slice with probability $1/pk$. Hence the number of items in the slice is distributed as the sum of n independent variables, each of which is 1 with probability $1/pk$ and 0 otherwise. Thus an application of Hoeffding's bound (Theorem 2.5, with $x = \epsilon\sqrt{n}/pk$) shows that

$$\Pr\{S_n \geq (n/pk)(1+\epsilon)\} = \Pr\{S_n - \mathsf{E}[S_n] \geq \epsilon n/pk\} \leq e^{-n\epsilon^2/2p^2k^2}.$$

Since there are pk slices, by Boole's inequality the probability that any of them has more than $(n/pk)(1+\epsilon)$ items is

$$q_n = \Pr\{\bar{A}\} \leq pk e^{-n\epsilon^2/2p^2k^2},$$

which is still exponentially small in n. Hence (2.41) becomes

$$\mathsf{E}[H(L_n)] \leq (1-q_n)g(n) + q_n g'(n) \sim g(n) = \frac{n}{p}\frac{k+1}{k}(1+\epsilon).$$

Since this holds for arbitrarily large k and arbitrarily small ϵ, we can combine it with (2.42) to obtain

$$\mathsf{E}[\mathrm{OPT}(L_n)] \sim n/p.$$

The general structure of this approach also appears in the proof of the upright matching theorem in Section 3.2 and in the analysis of two-dimensional packing in Section 7.1.1.

2.6 Monotonicity

We say that a bin-packing algorithm is *monotonic* if increasing the size of some of the items, or adding new items, can never decrease the number of bins used by the algorithm. (For the scheduling problem, an analogous monotonicity property could be defined, but it does not appear to be as useful in that problem.) OPT is clearly monotonic. Of the algorithms discussed in Section 1.4.2, one can show that NF is monotonic, but that the FF and FFD algorithms and the BF and BFD algorithms are not (see [John73, Murg88]).

As an example of the use of monotonicity, suppose we start with a list $L_n = (X_1, X_2, \ldots, X_n)$, and create a new list $L'_n = (X'_1, X'_2, \ldots, X'_n)$, where

2.6. MONOTONICITY

$X'_n = r\lceil X_n/r \rceil$ for some constant $r > 1$. Informally, we are rounding each element up to a multiple of $1/r$. Monotonicity assures us that $\mathrm{OPT}(L_n) \leq \mathrm{OPT}(L'_n)$. As we will see below, by a somewhat different rounding technique we can make the difference between the solutions quite small. The fact that L'_n has only r distinct item sizes can be very helpful. For example, it can enable us to use the linear programming approximation, which we now describe.

The bin-packing problem can be viewed as a type of integer program. By relaxing this to a linear program, we can gain some insight into the problem. We want to consider problems in which many items have the same size, so we will define somewhat different notation in this section. Suppose there are N different sizes, $s_1, s_2, \ldots, s_N \in (0, 1]$, and we have m_j items of size s_j; for reasons that will become clear, we will later, when we relax the integer program to a linear program, allow the possibility that the m_j be nonintegers. If $\vec{c} = (c_1, c_2, \ldots, c_N)$, where $0 \leq c_j$, satisfies $\sum_{j=1}^{N} c_j s_j \leq 1$, we will say that $\vec{c}$ is a possible *configuration* for a bin. Let $\mathcal{C}$ be the set of all possible such configurations. (Note that $|\mathcal{C}|$ may be a very large number; in fact, since we have not given any positive lower bound on the s_j, we cannot give any bound on $|\mathcal{C}|$ as a function only of N. However, since each s_j is positive, we know that $|\mathcal{C}|$ is finite.) Let $M = |\mathcal{C}|$, and for each $0 \leq j \leq N$ and $1 \leq k \leq M$, let C_{kj} be the number of items of size s_j in the kth configuration. Now to perform the packing of all the bins, we only need to decide how many times to use each configuration; let t_k be the number of times we use the kth configuration. We obtain the following integer program I:

$$\text{minimize} \quad \sum_{k=1}^{M} t_k$$

$$\text{subject to} \quad \sum_{k=1}^{M} t_k C_{kj} \geq m_j \quad \text{for } j = 1, \ldots, N,$$

$$t_k \geq 0 \quad \text{for } k = 1, \ldots, M. \quad (2.43)$$

(Note that in the middle line of (2.43) we could replace "$\geq m_j$" by "$= m_j$" without changing the value of the solution.) The solution to this integer program will yield the optimum number $\mathrm{OPT}(I)$ of bins needed to pack the items. Let $\mathrm{LIN}(I)$ be the solution to (2.43) with the constraints on the t_k relaxed so that they may be arbitrary nonnegative reals.

It is of interest to see the relationship between these two problems. Clearly, relaxing the constraints can only lower the minimum, so $\mathrm{LIN}(I) \leq \mathrm{OPT}(I)$. To obtain a bound in the other direction, we may solve $\mathrm{LIN}(I)$ and then round each t_k up to the next integer. Although the number of t_k may be very large, we know that there exists an optimum solution in which the number of nonzero t_k is at most equal to the number N of constraints, so we have the following.

Lemma 2.8 $\text{LIN}(I) \leq \text{OPT}(I) \leq \text{LIN}(I) + N$.

In Section 2.7.2 we will give an application of this bound. In [CW86a] the notion of configurations and linear algebra are used to give a characterization of the growth rate of wasted space in a packing for discrete distributions; a further discussion can be found at the end of Section 5.3.

2.7 More specialized techniques

Classical results from applied probability have been used in a variety of ways, some involving considerable ingenuity. We describe several of these results in this section.

2.7.1 Applications of the Poisson process

As we mentioned in Chapter 1, the lack of independence can be a significant source of difficulty. Occasionally, one can produce the desired independence by a further randomization of the input. For example, let N be a random variable with a Poisson distribution and mean value n. Equivalently, N may be taken as the number of renewals in [0,1] of a Poisson process with rate parameter n. If we draw N i.i.d. values from $U(0,1)$, then for any interval $I \subseteq [0,1]$ the number of points in I and the distribution thereof are independent of events outside of I. The standard deviation of N is $\sqrt{n}$, and for large n, N/n is likely to be very near 1, so in some sense the distribution just mentioned is not far from the distribution of n i.i.d. values drawn from $U(0,1)$. In some cases one can thus use the Poisson distribution to simplify an analysis. An application of this approach to a packing problem appears in Section 6.1.1.

A useful, complementary view of the above property of Poisson processes can be stated as follows (see [Çinl75], for example). Let $0 < t_1 < t_2 < \cdots < t_{n+1}$ be the first $n+1$ event times of a Poisson process with rate parameter 1, and define $t_0 = 0$. By definition, the differences $t_{i+1} - t_i$, $i = 1, 2, \ldots, n$, are i.i.d. with an exponential distribution. It is well known that the ratios $t_i/t_{n+1}, 1 \leq i \leq n$, are independent of t_{n+1} and distributed as the order statistics of n i.i.d. uniform random draws from [0,1]. Thus

2.7. MORE SPECIALIZED TECHNIQUES

in some cases replacing these order statistics by the t_i can be useful when analyzing problems involving the differences of these order statistics. (See [KT81, Chapter 13] for more discussion of the relationship between the Poisson process and order statistics.)

As an example of the power of this observation to simplify analyses, consider the following variation of the LPT scheduling rule discussed in Section 1.4.1, which we call PLPT for *Paired Largest Processing Time first*. (The analysis below can be found in [CFL84b], where PLPT is motivated by the easily proved fact that a reversal of the schedule leads to minimum total flow time.) Again we begin by sorting the tasks into order of decreasing size. Now, however, we schedule tasks two at a time; specifically, we take the largest two remaining tasks and place one onto each processor, with the larger of the two going to the processor having the smaller workload so far. If the number of tasks is odd, the last (smallest) task to be scheduled goes onto the processor with the smaller workload. Thus, if we let V_i be the variation between processor workloads just after tasks $X_{(n)}, X_{(n-1)}, \ldots, X_{(i+1)}$ have been scheduled, we have

$$V_n = 0,$$
$$V_i = |V_{i+2} - (X_{(i+2)} - X_{(i+1)})| \quad \text{for } n - i \text{ even},$$
$$V_0 = |V_1 - X_{(1)}| \quad \text{if } n \text{ is odd}. \tag{2.44}$$

(We leave V_i undefined if $i \neq 0$ and $n - i$ is odd.) The distribution of V_0 may not be immediately apparent from this recurrence. To see what this distribution is, first, let $\hat{X}_{(1)}, \hat{X}_{(2)}, \ldots$ be the successive epochs of a Poisson process with rate parameter 1; i.e., $\hat{X}_{(i+1)} - \hat{X}_{(i)}$, $i = 0, 1, 2, \ldots$, with $\hat{X}_{(0)} = 0$, are i.i.d. random variables having an exponential distribution with mean 1. We may define corresponding variables $\hat{V}_i$ by

$$\hat{V}_n = 0,$$
$$\hat{V}_i = |\hat{V}_{i+2} - (\hat{X}_{(i+2)} - \hat{X}_{(i+1)})| \quad \text{for } n - i \text{ even},$$
$$\hat{V}_0 = |\hat{V}_1 - \hat{X}_{(1)}| \quad \text{if } n \text{ is odd}.$$

Given $\hat{X}_{(n+1)}$, the random variables $\hat{X}_{(i)}/\hat{X}_{(n+1)}$ are distributed as the order statistics of n independent samples from $U(0, 1)$, so given $\hat{X}_{(n+1)} = x$, we have

$$\mathsf{E}[\hat{V}_0 \mid \hat{X}_{(n+1)} = x] = x\mathsf{E}[V_0].$$

Since $\mathsf{E}[\hat{X}_{(n+1)}] = n + 1$, we have

$$\mathsf{E}[\hat{V}_0] = (n+1)\mathsf{E}[V_0]. \tag{2.45}$$

Thus it remains to determine $\mathsf{E}[\hat{V}_0]$; this, however, is straightforward. By construction, the distributions of $\hat{X}_{(1)}$ and of each difference $\hat{X}_{(i+1)} - \hat{X}_{(i)}$ are independent and exponential with mean 1. Then by a simple induction, each $\hat{V}_i$ has an exponential distribution with mean 1, since the (absolute) difference between two independent exponentially distributed random variables with mean 1 is again exponentially distributed with mean 1 (see [Fell71, Section II.4]). In particular, $\mathsf{E}[\hat{V}_0] = 1$. Thus by (2.45) we see that the expected final variation in the original problem (solving (2.44)) is

$$\mathsf{E}[V_0] = \mathsf{E}[\hat{V}_0]/(n+1) = 1/(n+1).$$

Another illustration of this technique is provided in Section 4.1.

We have noted a tight relationship between the exponential distribution and the differences of order statistics of uniform random variables. It is also worth noting that the differences of order statistics of i.i.d. exponentially distributed random variables can be easily expressed.

Lemma 2.9 ([Fell71, Section I.6, Example (a)]) *Let $X_{(1)} \leq X_{(2)} \leq \cdots \leq X_{(n)}$ be the order statistics of n i.i.d. exponential random variables with mean 1, and define $X_{(0)} = 0$. Then the differences $X_{(i+1)} - X_{(i)}$, $0 \leq i \leq n-1$, are independent, and $X_{(i+1)} - X_{(i)}$ is exponentially distributed with mean $1/(n-i)$.*

An application of this lemma will appear in Section 4.2.

2.7.2 Kolmogorov-Smirnov statistics

Let $X_1, X_2, \ldots, X_n$ be a set of i.i.d. samples according to some *continuous* distribution F. Their *sample distribution function* is defined as

$$F_n(x) = \frac{1}{n}\big|\{i : X_i \leq x\}\big|.$$

For large n we would expect that the difference between F_n and F would be small. The Kolmogorov-Smirnov statistic, defined as

$$D_n = \sup_x \big|F_n(x) - F(x)\big|,$$

provides a measure of this difference and has been studied extensively (see [Durb73, Chapters 2–3]). Of special interest here will be the *one-sided Kolomogorov-Smirnov statistics* given by

$$D_n^+ = \sup_x \big(F_n(x) - F(x)\big),$$

2.7. MORE SPECIALIZED TECHNIQUES

$$D_n^- = \sup_x \left(F(x) - F_n(x) \right).$$

It is not hard to show that these statistics both have the same distribution, and that this distribution does not depend on the distribution F (assuming F is continuous); in fact, the distribution of D_n^+ and D_n^- is the same as that of

$$\max_{1 \le j \le n} \left\{ \frac{j}{n} - U_{(j)} \right\},$$

where the $U_{(j)}$ are the order statistics of n i.i.d. samples from $U(0,1)$. Smirnov has shown that (see [Durb73, Section 3.6])

$$\Pr\left\{ D_n^+ \le \frac{x}{\sqrt{n}} \right\} = 1 - e^{-2x^2}\left(1 - \frac{2x}{3\sqrt{n}} + O(1/n)\right). \tag{2.46}$$

(The constants hidden by the O-notation are independent of n but *not* independent of x.) It is also easily established that

$$\Pr\left\{ D_n^+ > \frac{x}{\sqrt{n}} \right\} = O\left(e^{-\alpha x^2}\right)$$

for some $\alpha > 0$. As a consequence we can easily compute

$$\mathsf{E}[nD_n^+] = \mathsf{E}[nD_n^-] = \Theta\left(\sqrt{n}\right). \tag{2.47}$$

In applications, the inequality $D_n^+ \le d$ can occasionally be transformed directly into a corresponding bound on a given performance measure. For example, Bruno and Downey [BD86] observed how this works in analyzing the bound in (2.40) when task sizes are independent samples from $U(0,1)$. In this case $D_n^+ \le d$ means that $i/n - X_{(i)} \le d$ holds for all $i = 1, \ldots, n$, so a summation yields $\frac{n+1}{2} - dn \le \sum_{i=1}^n X_i$. Substituting into (2.40) gives

$$R^{\mathrm{LS}}(L_n, m) \le \frac{m-1}{(n+1)/2 - dn} \le \frac{2(m-1)}{n} \frac{1}{1-2d}.$$

Then the estimate (2.46) can be used to bound the distribution of $R^{\mathrm{LS}}(L_n, m)$. Some applications of Kolmogorov-Smirnov statistics to the analysis of bin-packing can be found in [BD85, Knöd81, RT89b, RT89c]; we make no claim that this list is exhaustive.

In a demonstration of the power of the estimate (2.47), suppose we have an arbitrary (not necessarily continuous) distribution F over $[0,1]$. As usual, let $\mathrm{OPT}(L_n)$ be the number of bins required to pack n items drawn according to F, and define the *packing constant* for F to be $c = \lim_{n \to \infty} \mathsf{E}[\mathrm{OPT}(L_n)]/n$.

(Informally, the packing constant is the limiting expected number of bins used per item in an optimum packing.) We will give a simple proof that $\mathsf{E}[\mathrm{OPT}(L_n) - nc]$ is $O(\sqrt{n})$. (This is actually a special case of a result in [RT89f], but it is of interest to see how easily it can be obtained by the use of the Smirnov estimate and the linear programming approximation of Lemma 2.8.)

To begin, we let m be a parameter to be chosen later and define a new distribution G consisting of m atoms, each of weight $1/m$, at $s_j = F^{-1}(j/m)$, for $j = 1, 2, \ldots, m$. (Here we define $F^{-1}(z) = \min\{x : F(x) \geq z\}$.) Let c_m be the packing constant for G. It is not hard to see that

$$c_m - 1/m \leq c \leq c_m,$$

so we investigate c_m. Note that generating n items X_i distributed according to G can be achieved by taking n uniform samples U_i and setting $X_i = F^{-1}(\lceil mU_i \rceil / m)$. Now

$$nD_n^- = \max_{0 \leq z \leq 1}\{nz - |\{i : U_i \leq z\}|\} = \max_{0 \leq z \leq 1}\{|\{i : U_i > z\}| - (1-z)n\}.$$

Clearly, if we remove the items generated by the largest nD_n^- of the U_i and pack them in individual bins, we will be left with a set of items X_i' with sizes in $\{s_j\}_{1 \leq j \leq m}$ such that

$$|\{i : X_i' > s_j\}| \leq n(1 - j/m) \quad \text{for } 1 \leq j \leq m. \tag{2.48}$$

Let $\mathrm{OPT}(I)$ be the number of bins required to pack these remaining items. Note that $\mathrm{OPT}(I) \leq \mathrm{OPT}(I')$, where I' contains exactly $\lceil n/m \rceil$ items of each size s_j. Finally, let $\mathrm{LIN}(I')$ be a linear programming relaxation of this problem, in which we are to pack exactly n/m (rather than $\lceil n/m \rceil$) of each item size. In view of Lemma 2.8, we have $\mathrm{OPT}(I') \leq \mathrm{LIN}(I') + m$. It is not hard to see from the law of large numbers that $\mathrm{LIN}(I') = nc_m$. Hence we can bound the total number $\mathrm{OPT}(L_n)$ of bins required to pack all of the items by

$$\mathrm{OPT}(L_n) \leq nD_n^- + m + nc_m \leq nD_n^- + m + n(c + 1/m).$$

(See [Rhee90] for an asymptotically much stronger bound.) Then taking expectations using (2.47), and letting $m = \sqrt{n}$, we obtain

$$\mathsf{E}[\mathrm{OPT}(L_n)] - nc = O(\sqrt{n}),$$

where the constants hidden by the O-notation are independent of the distribution.

2.7.3 The second moment method

By an application of Chebyshev's inequality it is not difficult to prove that for a random variable whose first and second moments exist,

$$\Pr\{Y = 0\} \leq \frac{\mathsf{E}[Y^2] - \mathsf{E}[Y]^2}{\mathsf{E}[Y]^2}.$$

This inequality has often been used to show the existence of combinatorial quantities; its use is called the *second moment method* (see [ES74, Chapter 16]).

L. A. Shepp[3] [Shep72a, Shep72b] has shown how one can prove a somewhat stronger inequality by a different route. Using Schwarz' inequality, we have

$$\mathsf{E}[Y]^2 = \mathsf{E}[1_{\{Y \neq 0\}} Y]^2 \leq \mathsf{E}[1_{\{Y \neq 0\}}] \mathsf{E}[Y^2] = \Pr\{Y \neq 0\} \mathsf{E}[Y^2],$$

so that

$$\Pr\{Y = 0\} \leq \frac{\mathsf{E}[Y^2] - \mathsf{E}[Y]^2}{\mathsf{E}[Y^2]}.$$

In Section 4.3 we will let Y be the number of feasible solutions to the partition problem that satisfy a certain constraint and use this result to show that Y is likely to be nonzero. This will enable us to state an asymptotic bound for the median of the optimum solution.

Recently Berger [Berg91] has presented a related technique called the *fourth moment method*. She notes that by a simple direct argument, or by Hölder's inequality, one can show that for an arbitrary random variable Y we have

$$\mathsf{E}[|Y|] \geq \frac{\mathsf{E}[Y^2]^{3/2}}{\mathsf{E}[Y^4]^{1/2}}. \tag{2.49}$$

Here we give a simple illustration of this method.[4] Let Y be the sum of n independent variables $X_1, X_2, \ldots, X_n$ each of which is $+1$ with probability p, -1 with probability p, and 0 with probability $1 - 2p$. Assume $n > 1$; p is allowed to vary with n provided $np \geq 1$. We wish to estimate $\mathsf{E}[|Y|]$. (Note that because we are letting p vary with n, the argument leading to (2.29) (page 23) may not hold here; in particular, Y need not approach a normal distribution as $n \to \infty$.)

---

[3]Shepp attributes the basic idea to [Bill65].
[4]Berger gives considerably deeper examples than the simple one presented here, by applying this inequality to the design of parallel algorithms.

Using the fourth moment method we can estimate $\mathsf{E}[|Y|]$ easily. One readily computes that $\mathsf{E}[Y^2] = np$. Now $\mathsf{E}[Y^4]$ can be expressed as

$$\sum_{i=1}^{n}\sum_{j=1}^{n}\sum_{k=1}^{n}\sum_{l=1}^{n} \mathsf{E}[X_i X_j X_k X_l].$$

Note that some of the indices i, j, k, and l may be equal. In fact, it is not hard to see that $\mathsf{E}[X_i X_j X_k X_l]$ vanishes unless all four indices are equal or two of them equal one value and the other two equal another value. Using this observation we can readily compute $\mathsf{E}[Y^4] = \Theta(n^2 p^2)$, and thus by (2.49) we have

$$\mathsf{E}[|Y|] = \Omega(\sqrt{np}),$$

where the hidden constants are independent of both n and p. Using Theorem 2.6 it is not hard to establish that this bound is tight, so

$$\mathsf{E}[|Y|] = \Theta(\sqrt{np}). \tag{2.50}$$

This fact will be useful in Section 3.1.1.

2.7.4 An application of renewal theory

Let $Y_1, Y_2, \ldots$ be a sequence of i.i.d. nonnegative random variables, and let S_n be the partial sums given by

$$S_n = \begin{cases} 0 & \text{for } n = 0, \\ S_{n-1} + Y_n & \text{for } n > 0. \end{cases}$$

For convenience, let Y be a random variable distributed as the Y_i. If we imagine that the Y_i are distributed as the lifetime of some component that is replaced when it fails, we can interpret S_n as the time until the nth failure. This process is called a *renewal process*, and the "failures" are called *renewal events*. Sometimes it is more useful to think of the process from a different point of view: we let N_t be the number of failures during the time period $[0, t]$, so that

$$\Pr\{N_t \geq n\} = \Pr\{S_n \leq t\}.$$

Assuming that Y has mean μ and variance σ^2, one can show that $\mathsf{E}[N_t] \sim t/\mu$, which fits well with the intuition that the failure rate is about one item every μ time units. In fact, one can prove a much stronger description of $\mathsf{E}[N_t]$. The following definition will be useful: if Y assumes only nonnegative integer values, we say that the process is periodic if there is an integer $k > 1$ such that $\Pr\{Y = i\} \neq 0$ only when i is a multiple of k; a process that is not periodic is *aperiodic*.

2.7. MORE SPECIALIZED TECHNIQUES

Theorem 2.10 *Suppose the time between renewal events in a renewal process is distributed as a random variable Y, with mean μ and variance σ^2. Then*

a) if Y assumes only nonnegative integer values, and the process is aperiodic, then [Fell68, Section XIII.12, Exercise 22]

$$E[N_t] = \frac{t}{\mu} + \frac{\sigma^2 + \mu - \mu^2}{2\mu^2} + o(1),$$

and

b) if Y has a density, then [Cox62, Section 4.2]

$$E[N_t] = \frac{t}{\mu} + \frac{\sigma^2 - \mu^2}{2\mu^2} + o(1).$$

In [CFGR], renewal theory is applied to a variation of the bin-packing problem in which we say a bin is *covered* if it is packed to a level of at least 1, and we wish to maximize the number of bins covered; let us call this the *bin-covering problem*. Next Fit is easily adapted to this new problem. We initially select an empty bin as the current bin. Then, for each item, we pack it in the current bin; if this causes the bin level to be at least 1, close this bin and start a new (empty) current bin. Call this algorithm *Covering Next Fit*, and define $\text{CNF}(L_n)$ to be the number of bins covered by this algorithm. (Note that at the end of the algorithm, the then current bin will not count as being covered, even if it contains some items, because the total size of these items will be less than 1.)

Assume that the items to be packed are i.i.d. uniform over $[0, 1]$. It is not hard to compute that

$$\Pr\left\{\sum_{i=1}^{k} X_i \leq x\right\} = \frac{x^k}{k!} \quad \text{provided } x \leq 1.$$

It is convenient to suppose temporarily that an infinite sequence of items is available to be packed. Then, if we let v_j denote the number of items in the covering of the jth bin, we see that

$$\Pr\{v_1 > k\} = \Pr\left\{\sum_{i=1}^{k} X_i < 1\right\} = \Pr\left\{\sum_{i=1}^{k} X_i \leq 1\right\} = \frac{1}{k!}. \quad (2.51)$$

Note that an item is always packed into the first bin for which it is considered, so the v_j are i.i.d. Hence the number of bins covered by n items corresponds to the number of renewal events in the first n time units of a renewal process in which the time between renewal events is distributed as given in (2.51). Hence, by Theorem 2.10, if we let μ and σ^2, respectively, represent the mean and variance of v_1, we have

$$\mathsf{E}[\mathrm{CNF}(L_n)] = \frac{n}{\mu} + \frac{\sigma^2 + \mu - \mu^2}{2\mu^2} + o(1). \tag{2.52}$$

The quantities μ and σ^2 are readily computed using generating functions. (For discussion of generating functions of random variables the reader is referred to [Fell68, Knut73, Luek80, PB85].) By (2.51) we have

$$\sum_{i=0}^{\infty} \Pr\{v_1 > k\} z^k = e^z,$$

so since $\Pr\{v_1 = k\} = \Pr\{v_1 > k-1\} - \Pr\{v_1 > k\}$, the generating function for the random variable v_1 is

$$P(z) = \sum_{i=0}^{\infty} \Pr\{v_1 = k\} z^k = 1 + (z-1)e^z.$$

Hence

$$\mu = P'(1) = z e^z \big|_{z=1} = e,$$

and

$$\sigma^2 = P''(1) + P'(1) - \mu^2 = (z+1)e^z \big|_{z=1} + z e^z \big|_{z=1} - e^2 = 3e - e^2.$$

Inserting these values into (2.52), we obtain the remarkably precise estimate

$$\mathsf{E}[\mathrm{CNF}(L_n)] = \frac{n+2}{e} - 1 + o(1).$$

(For further investigations of the probabilistic behavior of the bin covering problem, see [RT89d, Rhee90].)

Chapter 3

Matching problems

Let n plus points and n minus points be chosen independently and uniformly at random in the unit square. Let M_n denote a matching of the plus points to minus points and let (P^-, P^+) denote a pair of matched points (or an edge) in M_n. Define $d(P^-, P^+)$ as the Euclidean distance (edge length) between P^- and P^+. In Section 3.1 we discuss the following estimate for the expected total distance between matched points, a result first proved by Ajtai, Komlós, and Tusnády [AKT84].

Theorem 3.1 *Let M_n denote a matching that minimizes the sum $\sum_{(P^-, P^+) \in M_n} d(P^-, P^+)$. Then*

$$\mathsf{E}\left[\sum_{(P^-, P^+) \in M_n} d(P^-, P^+)\right] = \Theta\left(\sqrt{n \log n}\right).$$

This total-edge-length matching problem is perhaps the most basic of the stochastic planar matching problems. Another problem, of even greater use to us, is the *up-right matching problem*. Here we have a total of n points, with each being an independent random draw from the unit square as before, and each being a plus or minus with equal probability. Thus the numbers of plus and minus points are described by a binomial distribution. Now let M_n denote a matching of plus points to minus points such that matched pluses are above and to the right of their corresponding minuses and such that the number of points so matched is a maximum. Thus, if $(P^-, P^+) \in M_n$, where $P^- = (x^-, y^-)$ and $P^+ = (x^+, y^+)$, then $x^- \leq x^+$ and $y^- \leq y^+$. The problem is to estimate the expected value of U_n, the number of points left unmatched in M_n. Figure 3.1 shows an example.

In [Shor86], Shor proved that $\mathsf{E}[U_n] = \Omega(\sqrt{n} \log^{3/4} n)$; also, Leighton and Shor [LS89] and Rhee and Talagrand [RT88a] showed that there exists a K

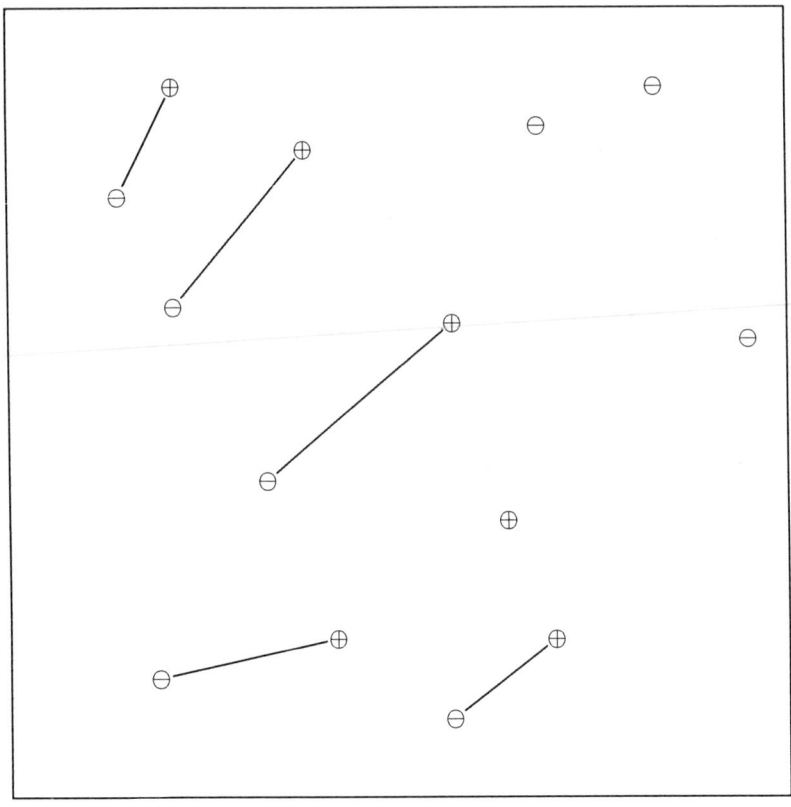

Figure 3.1: A maximum up-right matching. Here $n = 14$ and $U_n = 4$.

such that $\Pr\{U_n > K\sqrt{n}\log^{3/4} n\} = n^{-\Omega(\sqrt{\log n})}$, from which it follows that $\mathsf{E}[U_n] = O(\sqrt{n}\log^{3/4} n)$. Hence the following precise asymptotic estimate is available; it will be proved in Section 3.2.

Theorem 3.2
$$\mathsf{E}[U_n] = \Theta(\sqrt{n}\log^{3/4} n).$$

We will also study a third variation involving rightward matching, where matched pluses are required to be to the right of their corresponding minuses. A variation not covered here considers the problem instances of Theorem 3.1 and asks for an estimate of the expected *maximum* edge length in a matching that minimizes this quantity [LL82].

The literature on stochastic planar matching is relatively recent. A survey on the subject and its connections with bin packing can be found in [Shor85b]; this survey also discusses the duals of the matching problems expressed as linear programs.

In addition to bin packing, many applications have been cited for stochastic matching results. These include transportation assignment problems, VLSI design, computer storage allocation, and hypothesis testing in statistics. However, bin-packing applications provided a special impetus to the research. Here, the seminal paper is that of Karp, Luby, and Marchetti-Spaccamela [KLMS84], who discovered the relationship between two-dimensional bin packing and up-right matching.

The remainder of this chapter focuses on stochastic matching results to be used in the bin-packing analysis of later chapters. The use of Theorem 3.1 is indirect; the proof in Section 3.1 introduces a construction essential to the proof of a fundamental lower bound on on-line bin packing. The up-right matching estimate proved in Section 3.2 will be exploited in Sections 6.2 and 7.1.3 in the analysis of efficient one- and two-dimensional bin-packing algorithms. As will be seen, the proofs to follow create many opportunities to apply the fundamentals of Chapter 2. (On a first reading, one may wish to read the theorem statements but not the proof details. An understanding of the proofs, which are rather arduous, is not needed in later chapters.)

3.1 Proofs for Euclidean and rightward matching

This section begins with a proof of the $\Omega(\sqrt{n\log n})$ lower bound of Theorem 3.1. We then discuss the upper-bound proof. The section concludes with a rightward matching result that makes use of the lower-bound proof for Theorem 3.1. Our general approach to the proof of Theorem 3.1 is similar to that in [AKT84]. However, our lower-bound proof uses more elementary methods, so the details differ significantly.[1]

3.1.1 The lower bound

As is common in probabilistic analyses, part of the strategy of the following proof is the avoidance of difficulties caused by a lack of independence. Central to the proof is the following simple idea: suppose X is an integer-valued random variable whose distribution is symmetric about 0; that is, for all integers i, $\Pr\{X = i\} = \Pr\{X = -i\}$. Now suppose we construct two random variables Y and Z from X as follows. Let $Y = |X|$. If $X = 0$, let Z be ± 1 according to a toss of a fair coin; otherwise, let Z be 1 if $X > 0$ and -1 if $X < 0$. Then it is easy to see that Y and Z are independent. We will refer to this simple fact below as *symmetry independence*.

In preparation, we need the following definitions. Let S be a square with sides of length s parallel to the x or y axis. The *triangular regions of S* are the four triangles bounded by the edges and diagonals of S; these regions are numbered I–IV in Figure 3.2. For points $(x,y) \in S$, $f(S;x,y)$ denotes the distance to the nearest boundary of S. Define $f(S;x,y) = 0$ for all $(x,y) \notin S$. Thus, for (x,y) in the square S of Figure 3.2,

$$f(S;x,y) = \begin{cases} x - x_0 & \text{if } (x,y) \in \text{I}, \\ x_0 + s - x & \text{if } (x,y) \in \text{II}, \\ y - y_0 & \text{if } (x,y) \in \text{III}, \\ y_0 + s - y & \text{if } (x,y) \in \text{IV}. \end{cases} \quad (3.1)$$

Note that $f(S;x,y)$ is continuous; it is also differentiable except on the edges and diagonals of S. The derivatives $\left(\frac{\partial f}{\partial x}, \frac{\partial f}{\partial y}\right)$ take on the constant values $(1,0)$, $(-1,0)$, $(0,1)$, and $(0,-1)$ throughout the interior of regions I, II, III, and IV, respectively. When possible, we will use the abbreviation $f(S) = f(S;x,y)$.

$\mathsf{E}[f(S)]$ denotes the expected value of the distance to the nearest boundary from a point (x,y) chosen uniformly at random in S. A simple calculation shows that, for the square in Figure 3.2,

$$\mathsf{E}[f(S)] = \frac{s}{6}. \quad (3.2)$$

---

[1] The proof presented in this section was worked out in collaboration with P. W. Shor. The underlying ideas can be found in Shor's Ph.D. thesis [Shor85b].

3.1. PROOFS FOR EUCLIDEAN AND RIGHTWARD MATCHING

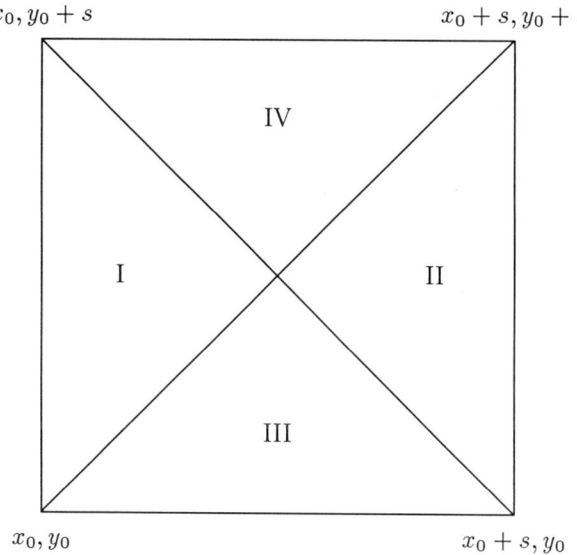

Figure 3.2: A square S and its triangular regions.

Let $\mathcal{P}$ be a set of points in the plane, each of which is labeled with a plus or minus. The subsets of plus and minus points are denoted by $\{P_l^+\}$, $\{P_l^-\}$, respectively. The *plus discrepancy* $\Delta^+(S)$ of a square S is simply the number of plus points of $\mathcal{P}$ in S minus the number of minus points of $\mathcal{P}$ in S.

In order to prove

$$\mathsf{E}\left[\sum_{(P^-,P^+)\in M_n} d(P^-, P^+)\right] = \Omega\left(\sqrt{n \log n}\right),$$

we construct a function $w = w(\mathcal{P})$, $\mathcal{P} = \{P_l^+\} \cup \{P_l^-\}$, such that[2]

$$\mathsf{E}\left[\sum_{j=1}^{n} w(P_l^+) - w(P_l^-)\right] = \Omega\left(\sqrt{n \log n}\right), \qquad (3.3)$$

and for any two points P and P' in the unit square

$$\frac{|w(P) - w(P')|}{d(P, P')} \leq 1. \qquad (3.4)$$

---

[2] As noted in [AKT84], this approach can be attributed to Ford and Fulkerson [FF62]. It is easy to verify that w arises as the dual function in a linear programming formulation of the matching problem.

To see how such a function would prove the lower bound, consider any matching of the plus points to the minus points, with the points indexed so that P_l^+ is matched to P_l^-, for $1 \le l \le n$. By (3.4) and then (3.3)

$$\mathsf{E}\left[\sum_{l=1}^n d(P_l^+, P_l^-)\right] \ge \mathsf{E}\left[\sum_{j=1}^n w(P_l^+) - w(P_l^-)\right] = \Omega\left(\sqrt{n \log n}\right),$$

and the lower bound is proved.

To develop the function w, we first define a simpler function v that satisfies (3.3) but not (3.4). Later, v will be modified so as to obtain a function w satisfying both (3.3) and (3.4).

Let $\mathcal{G}_i$, $i \ge 1$, denote the $4^i \times 4^i$ subdivision of the unit square into 16^i grid squares, each having sides of length 4^{-i}. Let S_{ij}, $j = 1, 2, \ldots, 16^i$ denote the jth grid square in $\mathcal{G}_i$. (How these squares are enumerated is unimportant.) $\mathcal{G}_0$ contains only the unit square S_{01}. Corresponding to the sequence $\mathcal{G}_0, \mathcal{G}_1, \ldots$, the function v is constructed in r stages according to the recurrence

$$v_0 = 0,$$
$$v_{i+1} = v_i + \delta_i \quad \text{for } i \ge 0, \tag{3.5}$$

with δ_i defined as follows.

For each j, divide up S_{ij} into its 16 constituent grid squares of $\mathcal{G}_{i+1}$, and then organize these grid squares into pairs as shown in Figure 3.3; paired squares are called *companions*. We want to define δ_i in (3.5) so that v_{i+1} is obtained from v_i by adding for each $j = 1, \ldots, 16^i$ either plus or minus $f(S_{i+1,j})/\sqrt{c \lg n}$ according as the plus discrepancy of $S_{i+1,j}$ is, respectively, greater or less than that of its companion. (Here, $c > 0$ is a universal constant whose role will be described later.) To accomplish this, put

$$\delta_i = \sum_{j=1}^{16^i} \delta_{ij}, \tag{3.6}$$

where

$$\delta_{ij} = \sum_{k=1}^{8} b(S_{ij}^{(k)})[f(S_{ij}^{(k)}) - f(\hat{S}_{ij}^{(k)})], \tag{3.7}$$

and for $1 \le k \le 8$,

$$b(S_{ij}^{(k)}) = \begin{cases} \frac{1}{\sqrt{c \lg n}} & \text{if } \Delta^+(S_{ij}^{(k)}) > \Delta^+(\hat{S}_{ij}^{(k)}), \\ \frac{-1}{\sqrt{c \lg n}} & \text{if } \Delta^+(S_{ij}^{(k)}) < \Delta^+(\hat{S}_{ij}^{(k)}). \end{cases} \tag{3.8}$$

3.1. PROOFS FOR EUCLIDEAN AND RIGHTWARD MATCHING   47

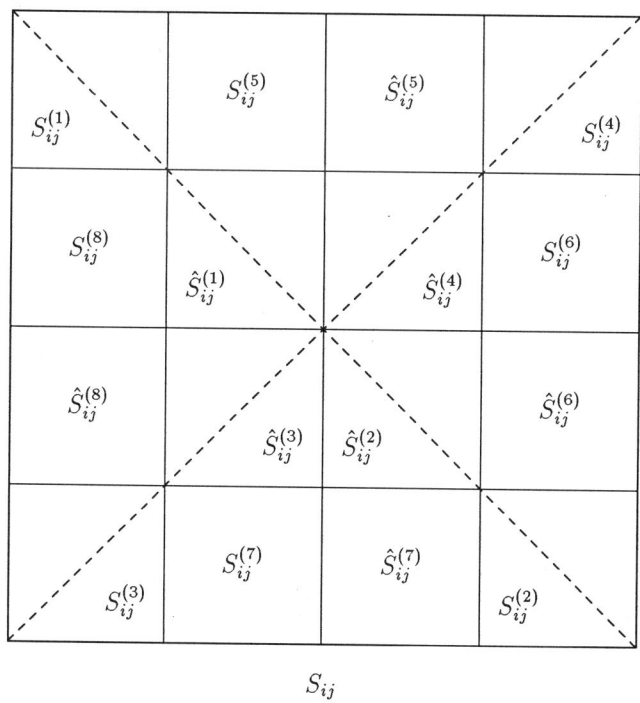

Figure 3.3: The pairs in S_{ij}. Note that the companion squares $S_{ij}^{(k)}$ and $\hat{S}_{ij}^{(k)}$ are either adjacent along a diagonal or they are the adjacent interior squares of the first or last row or column.

For resolving ties (i.e., $\Delta^+(S_{ij}^{(k)}) = \Delta^+(\hat{S}_{ij}^{(k)})$) we assign $b(S_{ij}^{(k)})$ one of the two values in (3.8) by the toss of a fair coin.

At first the reader might think that it would be simpler to base the sign choice in (3.8) on the sign of the plus discrepancy of a single square, rather than on the difference between the plus discrepancies of two companions. The advantage of the latter approach is that even though the numbers of plus points and minus points in S_{ij} may be conditioned (because of earlier stages in the process), the difference $\Delta^+(S_{ij}^{(k)}) - \Delta^+(\hat{S}_{ij}^{(k)})$ has a distribution symmetric about 0. This will enable us to take advantage of symmetry independence.

The desired function $v = v_r$ is is now obtained by putting $r = \lfloor \frac{1}{4} \lg n \rfloor$. For simplicity, we assume hereafter that n is such that $\frac{1}{4} \lg n$ is a positive integer. Note that then there is an average of one plus point and one minus point per grid square in $\mathcal{G}_r$. It is routine to extend the following analysis to general values of n; only the constants hidden in our asymptotic results change.

To show that v satisfies (3.3), we first fix $i \geq 0$ and verify that, for the function added at the $(i+1)$th stage,

$$\mathsf{E}\left[\sum_{l=1}^{n} \delta_i(P_l^+) - \delta_i(P_l^-)\right] = \Omega\left(\sqrt{\frac{n}{\log n}}\right). \quad (3.9)$$

Since this estimate is independent of i, the expected contribution of all $\frac{1}{4} \lg n$ stages gives the desired result in (3.3) with w replaced by v.

A bit of compact notation will be useful. Let

$$\sum_{l=1}^{n} f(S) - f(\hat{S}) \Big|_{P_l^-}^{P_l^+} =$$

$$\sum_{l=1}^{n} \left(f(S; P_l^+) - f(\hat{S}; P_l^+)\right) - \sum_{l=1}^{n} \left(f(S; P_l^-) - f(\hat{S}; P_l^-)\right).$$

By (3.6)–(3.8), the basic quantity to be estimated in a proof of (3.9) is $\mathsf{E}[b(S) \sum_{l=1}^{n} f(S) - f(\hat{S}) \big|_{P_l^-}^{P_l^+}]$, where S and $\hat{S}$ are companions in $\mathcal{G}_{i+1}$. The points of $\mathcal{P}$ in a grid square of $\mathcal{G}_{i+1}$ are distributed uniformly at random over that square. Let N^+ and N^- be the number of plus points and minus points in S, respectively. Define $\hat{N}^+$ and $\hat{N}^-$ similarly for $\hat{S}$. Then by (3.2)

3.1. PROOFS FOR EUCLIDEAN AND RIGHTWARD MATCHING

and (3.8)

$$\mathsf{E}\left[b(S)\sum_{l=1}^{n} f(S) - f(\hat{S})\,|_{P_l^-}^{P_l^+} \;\bigg|\; N^+, N^-, \hat{N}^+, \hat{N}^-\right]$$

$$= \frac{1}{6\sqrt{c\lg n}} \frac{1}{4^{i+1}} \mathsf{E}[|N^+ - N^- - (\hat{N}^+ - \hat{N}^-)|]. \tag{3.10}$$

Let A be the area of either of the two companions. By symmetry independence, the distribution of $|N^+ - N^- - (\hat{N}^+ - \hat{N}^-)|$ is not conditioned by any comparisons we have made thus far; its distribution is the same as that of the absolute value of a variable constructed by adding n variables, each of which is $+1$ with probability A, -1 with probability A, and 0 otherwise. From (2.50) (page 38) we know that $\mathsf{E}[|N^+ - N^- - (\hat{N}^+ - \hat{N}^-)|]$ is $\Theta(\sqrt{nA})$. Since in our case $A = 16^{-(i+1)}$, this expectation is $\Theta(\sqrt{n}/4^i)$. Thus by (3.10) the expected contribution from a single pair of companions is $\Theta(16^{-i}\sqrt{n/\log n})$. Since there are $\Theta(16^i)$ pairs of companions, the total contribution from them all is $\Theta(\sqrt{n/\log n})$, and (3.9) is established. Thus, if we had let $w = v$, we would have satisfied (3.3).

Next, we see how to modify the construction to ensure that (3.4) holds. We begin by analyzing the partial-derivative process $\{\frac{\partial v_i}{\partial x}\}_{0 \le i \le r}$, with $v_i = v_i(P)$ and P a point in some S_{rj} not on an edge or diagonal of S_{rj}. By symmetry a similar analysis applies to $\{\frac{\partial v_i}{\partial y}\}_{0 \le i \le r}$. The goal of the analysis is a proof of the following result, from which a proof of the lower bound follows readily.

Claim 3.3 *Let S and $\hat{S}$ be companions in $\mathcal{G}_r$. Then throughout the triangular regions of $S \cup \hat{S}$, we have $\frac{\partial v_i}{\partial x}, \frac{\partial v_i}{\partial y} \le \sqrt{1/2}$, $i = 1, \ldots, r$, with probability $1 - O(e^{-c/4})$.*

Proof of claim. First, from (3.1) and (3.8), observe that at point P we have $\frac{\partial v_{i+1}}{\partial x} = \frac{\partial v_i}{\partial x} \pm a$, where $a = 1/\sqrt{c\lg n}$ or 0. (Note that whether or not a is 0 depends only on the location of the point P.) It is convenient to retain the expanded set of sample paths in which stages adding or subtracting 0 are kept distinct. Note that the values $\pm 1/\sqrt{c\lg n}$ are equally likely at each stage where $a \ne 0$. That is, $\{\frac{\partial v_i}{\partial x}\}$ is a martingale sequence. This follows because, given the number of points in S_{ij}, for any two grid squares S and S' within S_{ij}, $\Delta(S) - \Delta(S')$ is equally likely to be positive or negative, independent of the plus discrepancy of S_{ij}. (Recall that we resolve ties by a flip of a fair coin.)

To estimate the maximum excursions of $\{\frac{\partial v_i}{\partial x}\}$, let $N_r \le r$ denote the number of nonzero steps in $\{\frac{\partial v_i}{\partial x}\}_{0 \le i \le r}$. Then for a given value of N_r, $\max_{0 \le i \le r} \left|\frac{\partial v_i}{\partial x}\right|$ is equal in distribution to $\max_{0 \le i \le N_r} |Z_i|$, where $\{Z_i\}$ is a symmetric random walk starting at the origin $Z_0 = 0$. It is easy to show that $\max_{0 \le i \le N_r} |Z_i|$ increases stochastically with N_r, so

$$\Pr\left\{\max_{0 \le i \le r} \left|\frac{\partial v_i}{\partial x}\right| > t\right\} \le \Pr\left\{\max_{0 \le i \le N_r} |Z_i| > t\right\}, \quad \text{for } t \ge 0. \quad (3.11)$$

If σ^2 denotes the variance of Z_r, then by Theorem 2.7 (page 20), suitably scaled, we have

$$\Pr\left\{\max_{0 \le i \le N_r} |Z_i| > \sigma t\right\} = O(e^{-t^2/8}).$$

Since $\sigma^2 = r/c \lg n = 1/4c$, the choice $\sigma t = \sqrt{1/2}$ leads to

$$\Pr\left\{\max_{0 \le i \le N_r} |Z_i| > \sqrt{1/2}\right\} = O(e^{-c/4}).$$

Then by (3.11) we have $\frac{\partial v_i}{\partial x} \le \sqrt{1/2}$, $i = 1, \ldots, r$, with probability $1 - O(e^{-c/4})$.

Now $\frac{\partial v_i}{\partial x}$ can take at most four distinct values in S_{rj}, depending on which triangular region of S_{rj} contains P. Thus, combining the above analysis with that of $\frac{\partial v_i}{\partial y}$ and extending the result on S_{rj} to both S_{rj} and its companion, applications of Boole's inequality prove the claim. ∎

We are now ready to modify v so as to obtain a new function w satisfying both (3.3) and (3.4). We need only change the definition of $b(S)$ in (3.8) so that, if $b(S)$ as defined would lead to a function such that $\left|\frac{\partial v_i}{\partial x}\right| > \sqrt{1/2}$ or $\left|\frac{\partial v_i}{\partial y}\right| > \sqrt{1/2}$ at some point in $S \cup \hat{S}$, then $b(S)$ is taken to be 0 instead. Then, by definition, we have for the new function $\frac{\partial w}{\partial x}, \frac{\partial w}{\partial y} \le \sqrt{1/2}$, and hence $\left(\frac{\partial w}{\partial x}\right)^2 + \left(\frac{\partial w}{\partial y}\right)^2 \le 1$ throughout the unit square. Property (3.4) follows directly.

For property (3.3) we note that, by symmetry independence, the event that a pair of companions is affected by the modification is independent of the value of $|N^+ - N^- - (\hat{N}^+ - \hat{N}^-)|$ for that pair. Thus by Claim 3.3, for large enough c, the contribution to (3.9) is affected only by a constant factor, so (3.3) will still hold. Thus both properties (3.3) and (3.4) hold and we are done. ∎

3.1.2 The upper bound

Only the lower bound result will be of later use. However, the upper-bound proof is worthy of discussion. Here we give a heuristic argument which is

3.1. PROOFS FOR EUCLIDEAN AND RIGHTWARD MATCHING

a slight variation on the proof in [AKT84].[3] The argument is constructive in that it defines an algorithm that produces a matching with the desired expected total edge length. We give below the construction. It is easy to argue heuristically that the algorithm gives a matching satisfying the $O(\sqrt{n \log n})$ bound; these remarks are given following the definition of the algorithm. For a proof of the upper bound see [AKT84], and for a completely different proof see [Tala91].

For convenience, we consider problem instances in which the number of points is a power of 4. Generally speaking, the algorithm shifts the positions of the points while successively refining the unit square into grids $\mathcal{G}'_i$ similar to those in the previous subsection. The $\mathcal{G}'_i$ are $2^i \times 2^i$ arrays of identical squares, so the earlier grids $\mathcal{G}_i$ are given by $\mathcal{G}_i = \mathcal{G}'_{2i}$ for $i \geq 0$. Each of the $r' = \frac{1}{2} \lg n$ steps of the algorithm consists of two stages; in step i the two stages are performed on each square in $\mathcal{G}'_{i-1}$. At the end of step i there will be exactly $n/4^i$ points in each grid square of $\mathcal{G}'_i$. The desired result is obtained from an estimate of the expected total distance between the initial and final positions of the points.

Figure 3.4 illustrates the process applied to the plus points of a square S in $\mathcal{G}'_{i-1}$. S has been translated to the origin for convenience; also assume that the right edge of S has x coordinate 1. The first stage begins by determining a *median line* dividing S vertically into two rectangles, each with half the points. The median line is placed midway between the closest points on its left and right. Now translate the median line to the midpoint of the square, at the same time "stretching" the x coordinates in the smaller rectangle and "shrinking" the x coordinates in the larger rectangle. The stretching and shrinking operations are only changes of scale that preserve the relative positions of the points within each half of S. For example, if the median line has x coordinate x_1, then a coordinate x in the smaller rectangle of Figure 3.4(a) is transformed to a new coordinate $x' = x/2x_1$ in Figure 3.4(b). The shrinking of coordinates in the right rectangle is defined symmetrically.

The second stage of step i is a repetition of the first, except that it is applied to both the left and right halves of S with the point distribution altered as in stage 1 and with the median lines now drawn horizontally so as to divide the rectangles into upper and lower rectangles with $n/4^i$ points each (see Figure 3.4(b)). Stage 2 concludes with the stretching and shrinking operations of stage 1 applied to each pair of rectangles.

---

[3]The authors acknowledge discussions with R. M. Karp, who attributes the present variation to J. Komlós.

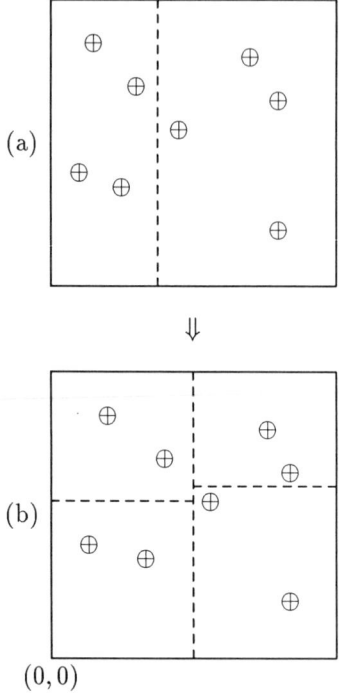

Figure 3.4: Moving plus points of $S \in \mathcal{G}_{i-1}$ in stage 1, step i. Part (a) shows the vertical median line at the beginning of stage 1. Note that in part (b) the portions of the square to the left and right of this line have been scaled horizontally so that they are of equal width; points are moved proportionately. Now the vertical median lines are drawn in both halves of the square, and in the next stage (not shown here) vertical scaling is done on each of the four subsquares so that both horizontal median lines divide the square equally.

3.1. PROOFS FOR EUCLIDEAN AND RIGHTWARD MATCHING

It is easy to see that at the end of the $r' = \frac{1}{2}\lg n$ steps there will be exactly one point in each of the n grid squares of $\mathcal{G}_{r'}$. The whole procedure is repeated on the minus points in S, so that we arrive at a distribution of points with exactly one plus point and one minus point in each grid square of $\mathcal{G}_{r'}$. Matching the points within such a grid square gives an average edge length of at most $\sqrt{2/n}$. Then matching all such pairs of points gives a matching with an $O(\sqrt{n})$ expected total edge length. Thus, since we have an $\Omega(\sqrt{n \log n})$ lower bound, an upper bound on the expected total Euclidean distance between the initial and final point positions produced by the r'-step transformation will yield an upper bound on the expected total edge length in a matching that minimizes the length.

It is easy to argue the expected-distance result heuristically. The processes on the plus points and minus points are obviously independent, as are the sequences of first stages and second stages in each process. Thus it is enough to describe the motion of the plus points in just one of the dimensions. By the successive rescalings of point positions it is natural to approximate the motion of a point in one of the coordinates by a symmetric random walk $\{X_i\}_{1 \leq i \leq r'}$, in which a point is equally likely to move left or right at each step, and the X_i are independent, bounded random variables with $\mathsf{E}[|X_i|] = O(\sqrt{1/n})$. Hence it is reasonable to suppose that the expected final displacement after $\frac{1}{2}\lg n$ steps is bounded by $O(\sqrt{\log n/n})$.

3.1.3 A rightward matching problem

Before getting into rightward matching we note some important extensions of Theorem 3.1. As is commonly the case with asymptotic results of this sort, the absence of information about multiplicative constants is compensated by a greater generality of the result. For example, it is not difficult to show that the $\Theta(\sqrt{n \log n})$ result also holds when [Shor86]

i) the number of points is Poisson distributed with mean n,

ii) the points are restricted to lie on an $n \times n$ lattice, or the problem is similarly discretized in just one of the dimensions, or

iii) the n plus points are fixed at the vertices of a $\sqrt{n} \times \sqrt{n}$ lattice and n minus points are chosen at random as before.

Similar extensions also apply to the up-right matching result in the next subsection.

The rightward matching extension here is more difficult to prove. However, the function developed in Section 3.1.1 supplies a useful tool for analyzing the following problem. The problem instance is as before, but the

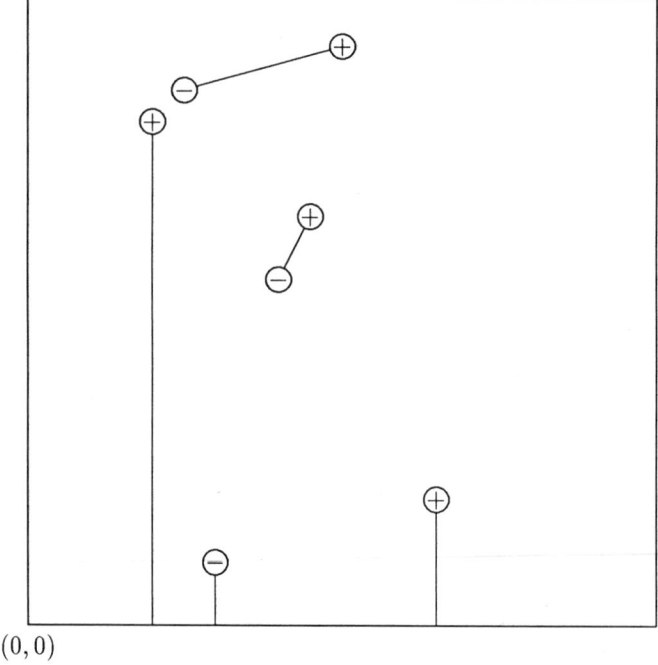

Figure 3.5: A rightward matching.

notion of matching is generalized as follows. In a valid matching, a point P can be matched either to a point of opposite sign or to a point on the lower boundary directly beneath P; in the latter case the edge length is the corresponding vertical distance. Whenever points of opposite sign are matched, the plus point must be to the right of the minus point.

Figure 3.5 gives an example. As illustrated in the figure, valid matchings do not require that the number of plus points matched to minus points be maximized. The matching characteristic of interest is the sum $V_n = V_n(M_n)$ of the vertical components of the edges in M_n. The reason for the special nature of this problem will be found in the bin-packing application of Section 6.2.

The following result was proved in [Shor86]; we adhere closely to Shor's proof.

Theorem 3.4
$$\mathsf{E}[V_n] = \Omega(\sqrt{n \log n}).$$

3.1. PROOFS FOR EUCLIDEAN AND RIGHTWARD MATCHING

Proof. The result will follow easily if we can produce a function v mapping the unit square into $[0,1]$ such that

1. $\frac{v(P)-v(P')}{d(P,P')} \leq 1$ for all points P, P' in the unit square,
2. $\mathsf{E}[\sum_{l=1}^{n} v(P_l^+) - v(P_l^-)] = \Omega(\sqrt{n \log n})$,
3. v is decreasing in x for $0 \leq x \leq 1$, and
4. $v = 0$ at the lower boundary of the unit square.

To see how such a function proves the theorem, let $Y(P)$ denote the y coordinate of an arbitrary point P, and let (P, P') denote a generic edge of a rightward matching M_n. Thus P denotes a minus point or P' denotes a plus point. If P' is a plus point, then P is either a minus point or a point on the lower boundary. Similarly, if P is a minus point, then P' is a plus point or a point on the lower boundary. Suppose, first, that $(P, P') = (P_i^-, P_j^+)$, and let (x_i^-, y_i^-) and (x_j^+, y_j^+) be the respective positions of P_i^- and P_j^+. Since M_n is a rightward matching, we have $x_i^- \leq x_j^+$. Thus, if Q is the point (x_i^-, y_j^+), then by property 3

$$v(P') - v(P) = v(P_j^+) - v(P_i^-) \leq v(Q) - v(P_i^-).$$

By property 1 we then have $v(P') - v(P) \leq d(Q, P_i^-) = |Y(P') - Y(P)|$. If either of P or P' is a point on the lower boundary, then both points lie on the same vertical line, so $v(P') - v(P) \leq d(P', P) = |Y(P') - Y(P)|$. Thus in either case

$$v(P') - v(P) \leq |Y(P') - Y(P)|. \tag{3.12}$$

Then using (3.12) and properties 4 and 2, we have

$$\mathsf{E}\left[\sum_{(P,P') \in M_n} |Y(P') - Y(P)|\right] \geq \mathsf{E}\left[\sum_{(P,P') \in M_n} v(P') - v(P)\right]$$

$$= \mathsf{E}\left[\sum_{l=1}^{n} v(P_l^+) - v(P_l^-)\right]$$

$$= \Omega(\sqrt{n \log n}).$$

It remains to develop a function v with properties 1–4. Note, first, that the procedure in Section 3.1.1 generates a function satisfying properties 1, 2, and 4, but not property 3. However, building on this procedure, we can construct the desired function v as follows.

First, take some interior horizontal slab R of the unit square, say, the middle third, $1/3 \leq y \leq 2/3$. By the procedure in Section 3.1.1 we now develop a function w determined by the subset of the instance $\mathcal{P}$ in R such that w has the properties

1'. $\frac{w(P)-w(P')}{d(P,P')} \le 1/6$ for all points P, P' in R,

2'. $\mathsf{E}[\sum_i w(P_{l_i}^+) - w(P_{l_i}^-)] = \Omega(\sqrt{n \log n})$, where $l_1, l_2, \ldots$ index the points in R, and

4'. $w = 0$ on the boundaries of R.

This requires only trivial modifications to the procedure in Section 3.1.1. Restriction to R is only a change of scale, and property 1' is obtained by changing some constants (see Claim 3.3). The effect of the modifications is clearly limited to the multiplicative constant hidden by the $\Omega(\sqrt{n \log n})$ result.

Finally, we define v in terms of the w above as follows.

$$v = \begin{cases} \frac{1}{2}(1-x)y & \text{if } 0 \le y \le 1/3, \\ w(x,y) + \frac{1}{6}(1-x) & \text{if } 1/3 \le y \le 2/3, \\ \frac{1}{2}(1-x)(1-y) & \text{if } 2/3 \le y \le 1. \end{cases} \quad (3.13)$$

(Note that the added terms have equal expectation on the plus and minus points.) By inspection of (3.13) and the properties of w, v has properties 1–4, so we are done. ∎

Quite recently, Shor [Shor90] and Talagrand [Tala91] have shown that the upper bound corresponding to Theorem 3.4 is $O(\sqrt{n \log n})$, thus giving a $\Theta(\sqrt{n \log n})$ result for rightward matching.

3.2 Proof of the up-right matching estimate

This section proves the $\mathsf{E}[U_n] = \Theta(\sqrt{n} \log^{3/4} n)$ result asserted by Theorem 3.2 for the up-right matching problem. A relatively compact proof of the $O(\sqrt{n} \log^{3/4} n)$ upper bound was recently presented by Coffman and Shor [CSb] and uses several of the ideas discussed in Chapter 2. The proof of the lower bound is given first, as it is much shorter; it is adapted from [Shor86] and uses ideas drawn from [AKT84]. Before getting into the proofs, we discuss a number of preliminary observations and lemmas.

It is convenient to reformulate our problem in terms of the discrepancy function,[4] introduced in Section 3.1.1. For any subset L of the unit square the plus discrepancy, $\Delta^+(L)$, is the number of plus points in L less the number of

---
[4]Problems of estimating discrepancies in various stochastic settings have received considerable attention from mathematicians; a brief history appears in [LS89, RT88a].

3.2. PROOF OF THE UP-RIGHT MATCHING ESTIMATE

minus points in L. The term discrepancy by itself refers to $\Delta(L) = |\Delta^+(L)|$. L is called a *lower layer* if it is closed (in the sense of point-set topology) and if $(x, y) \in L$ implies $(x', y') \in L$ whenever $x' \leq x$ and $y' \leq y$. A straightforward application of Hall's matching theorem (for the statement of Hall's matching theorem, see [BM76, Section 5.2] or [Manb89, Section 10.2.2]) shows that U_n has the same distribution as $\sup_{L \in \mathcal{L}} \Delta^+(L)$, where $\mathcal{L}$ is the set of all lower layers. We will prove

$$\mathsf{E}\left[\sup_{L \in \mathcal{L}} \Delta(L)\right] = \Theta(\sqrt{n} \log^{3/4} n); \tag{3.14}$$

the desired result follows trivially.

For each lower layer L there exist lower layers L' such that $\Delta(L) = \Delta(L')$ with probability 1 and such that the boundaries of L' are the unit intervals on the x and y axes and a third, nonincreasing boundary extending from $(0,1)$ to $(1,0)$. This third boundary is called a *lower-layer function*.

It is readily verified that the set of partitions of S created by lower-layer functions is also created by the subset consisting only of decreasing *step* functions. Now rotate the unit square 45° counterclockwise, center it at the point $(1/2, 0)$, and scale it down by a factor of $\sqrt{2}$. The problem instance changes as illustrated in Figure 3.6, where a lower-layer step function becomes a piecewise linear function $f(x)$, $0 \leq x \leq 1$, with the slopes of the pieces alternating between $+1$ and -1.

Hereafter, our terminology refers to this transformed version of the problem. Lower-layer functions are defined on $[0,1]$, they are completely contained in the rotated square, and they vanish at $x = 0$ and 1. Let $\mathcal{F}$ denote the subset of piecewise linear lower-layer functions with slopes alternating between -1 and $+1$.

3.2.1 The lower bound

Let N be a random variable having the Poisson distribution with mean n. We begin by noting that we can assume that the number of points is N, rather than n, without changing the expected number of unmatched points by more than $O(\sqrt{n})$; to see this, note that $\mathsf{E}[|n - N|] = O(\sqrt{n})$, and adding or deleting a single point can change the number of unmatched points by at most 1. The use of this Poisson distribution makes events involving points in disjoint regions of the square become independent; see Section 2.7.1. Thus we consider a random sample of N +'s and −'s in the rotated square. We then define a sequence $f_0, f_1, \ldots, f_k$ of successively more refined lower-layer

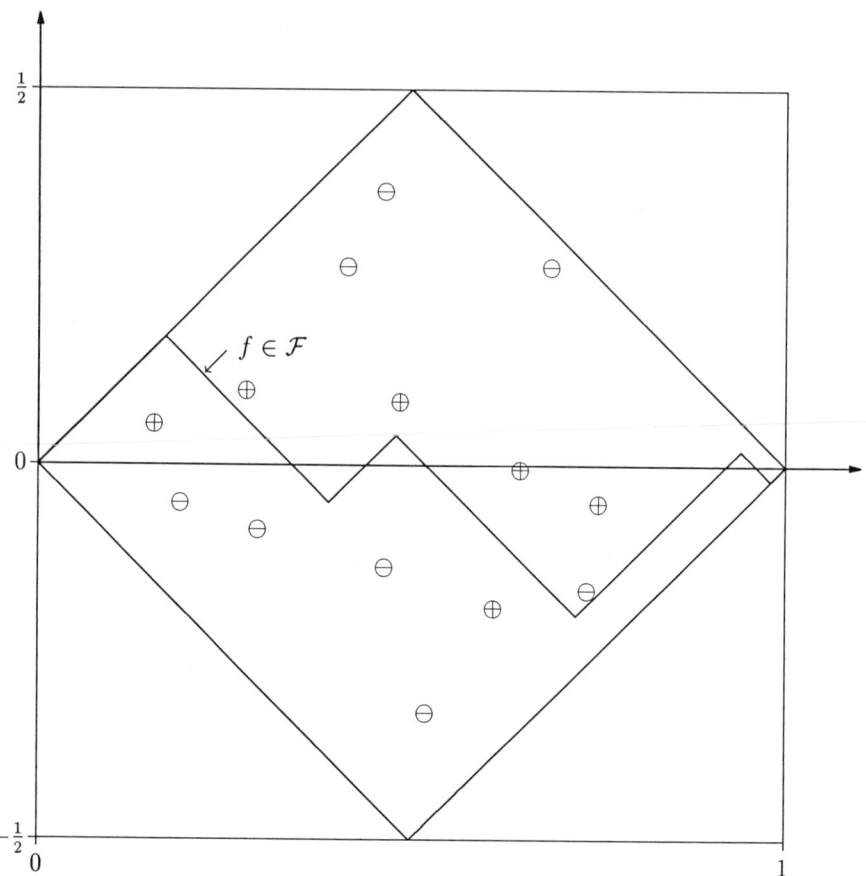

Figure 3.6: The transformed problem. The plus and minus points shown here are for the same data as in Figure 3.1. An example of a lower-layer function $f \in \mathcal{F}$ is also shown.

3.2. PROOF OF THE UP-RIGHT MATCHING ESTIMATE

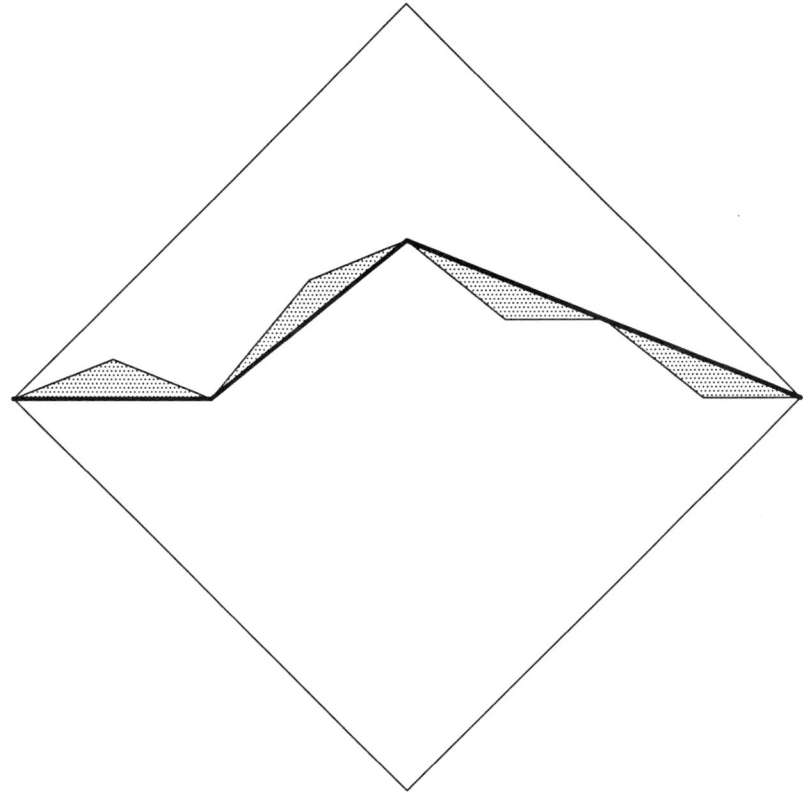

Figure 3.7: The sequence of triangles after a stage of the refinement process. The bold line is a lower-layer function.

functions. We show that the cumulative effect of successive refinements is a function $f = f_k$ with an expected plus discrepancy of $\Omega(\sqrt{n}\log^{3/4} n)$. The f_i are determined in sequence by a k-stage procedure that produces after each stage a sequence of adjacent triangles. Let the *base* of a triangle be its longest side. As illustrated in Figure 3.7, f_i is the piecewise linear function defined by bases of the triangles created by the first i stages.

The first stage begins with the single isosceles triangle shown in Figure 3.8, with vertices at $(0,0)$ and $(0,1)$ and with the shorter sides having slopes $\pm s$, where $s = 1/\sqrt{\lg n}$. A triangle T in the sequence produced at stage i has the following property. Each f_j, $j \geq i$, will pass through two vertices of T, and the portion of f_j between these vertices will be contained in T. For

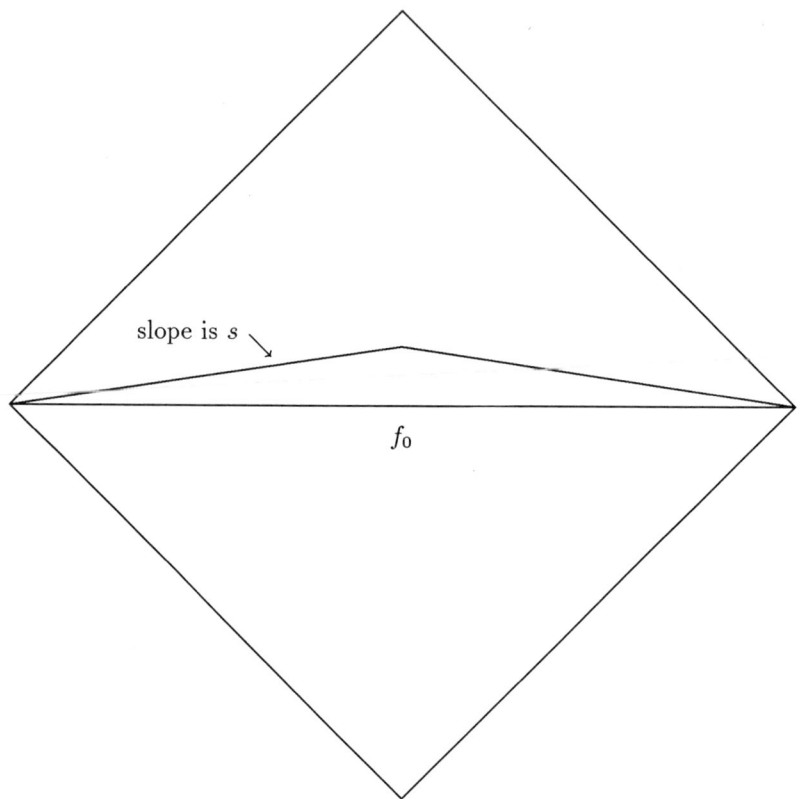

Figure 3.8: The initial triangle.

3.2. PROOF OF THE UP-RIGHT MATCHING ESTIMATE

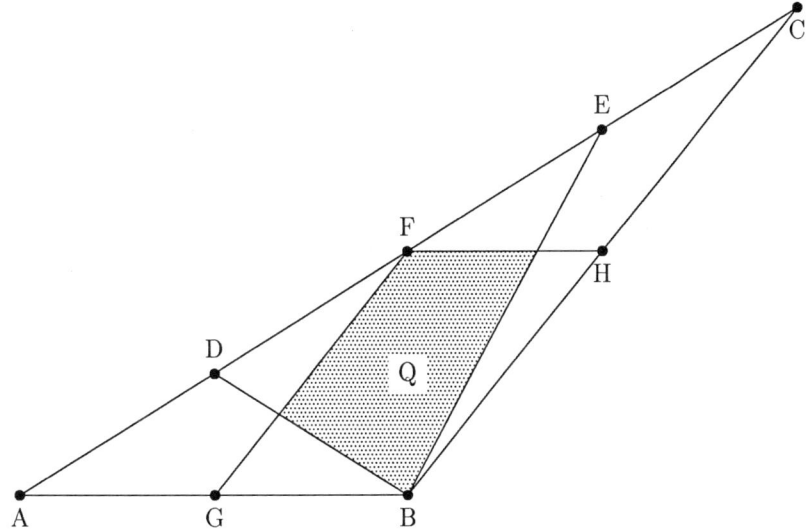

Figure 3.9: A typical triangle T.

example, in Figure 3.7, the final function f will lie in the shaded triangles. Each triangle T after the ith stage is processed in the $(i+1)$th stage so that T is replaced by two triangles, each with one-quarter of the area of T. The number of successive refinements is chosen to be $k = \lceil \lg n \rceil$ so that the smallest possible triangles after the final stage contain at most one point each on the average.

The procedure for refining a triangle T at stage i is illustrated in Figure 3.9. Points D, F, and E divide AC into quarters, G bisects AB and H bisects BC. In the sequence after stage i, T will be replaced by either triangles ADB and BEC or triangles AGF and FHC according to the plus discrepancy $\Delta^+(Q)$ of the central quadrilateral Q (shaded in the figure). If $\Delta^+(Q) < 0$, then the former choice is made so as to put Q outside the next lower layer (above f_i); if $\Delta^+(Q) > 0$, the latter choice is made and Q is included in the next lower layer. In the event that $\Delta^+(Q) = 0$, the choice is made according to a flip of a fair coin. Thus in this refinement Q contributes an expected increase of $\mathsf{E}[\Delta(Q)] = \mathsf{E}[|\Delta^+(Q)|]$ to the plus discrepancy of the lower layer.

In general, the central vertex B can point down, as in the figure, or up. In the latter case, the criterion for including Q is reversed; triangles ADB and BEC are selected if $\Delta^+(Q) > 0$, triangles AGF and FHC are selected if $\Delta^+(Q) < 0$, and a coin flip decides when $\Delta^+(Q) = 0$. Q's expected contribution then remains as above.

We call the iteration of this process for k stages the *full refinement procedure*. (In the final sequence some triangles will have base slopes with magnitudes exceeding 1; we will describe a truncated procedure below which addresses this problem.) We number the stages so that the initial triangle shown in Figure 3.8 is considered the result of stage 0.

For convenience in the proof, we will associate the triangles constructed as above with nodes in a perfect binary tree of height k, which we call the *full refinement tree*. The root of the tree is labeled with the triangle present at the beginning of the process, shown in Figure 3.9. Now suppose some triangle ABC, associated with node x in the tree, is replaced by ADB and BEC; then triangle ADB is associated with the left child of x, and BEC is associated with the right child of x. Similarly, if ABC is replaced by AGF and FHC, then triangle AGF is associated with the left child of x, and FHC is associated with the right child of x.

We now state four easily proved properties of the triangles created by the procedure. The properties are keyed to Figure 3.9. The first three can be proved by elementary geometric arguments; the fourth requires an induction along with the fact that the sum of the slopes of AD and DB is twice the slope of AB. Let the *base slope* of a triangle be the slope of its longest side.

1. The three vertices of T have x coordinates equally spaced, so B and F fall on a vertical line.
2. The two triangles created from T each have $1/4$ the area of T.
3. The area of the central quadrilateral is $1/3$ that of T.
4. If the base slope of T is is, then the other two sides of T have slopes $(i-1)s$ and $(i+1)s$. Moreover, the two triangles created from T are either both similar to T (e.g., AGF and FHC in Figure 3.9), or one has base slope $(i-1)s$ and the other has base slope $(i+1)s$ (e.g., ADB and BEC, respectively, in Figure 3.9).

It is easily verified from the properties listed above that at stage i the full refinement procedure tests 2^i central quadrilaterals, each of area $A = 4^{-i}s/6$, to determine whether they should be included in the next lower layer. Then an easy calculation shows that the inclusion or exclusion of a quadrilateral Q at stage i contributes an expected increase in the lower-layer plus discrepancy that is equal to $\mathsf{E}[\Delta(Q)] = \Theta(\sqrt{nA}) = \Theta(2^{-i}\sqrt{ns})$. Thus stage i contributes a total expected increase of $\Theta(\sqrt{n}\log^{-1/4} n)$. The quadrilaterals generated by the procedure are all disjoint, so the effect of all k stages is a lower layer with the desired plus discrepancy $\Theta(\sqrt{n}\log^{3/4} n)$.

3.2. PROOF OF THE UP-RIGHT MATCHING ESTIMATE

We now return to the problem mentioned above: we need to avoid the creation of triangles with base slopes of magnitude exceeding 1. This can be done quite simply by stopping the refinement process whenever it would produce a triangle that has a base slope outside of $[-1, 1]$; we continue to refine other triangles. Thus after each stage we have a function f_i that is a valid lower-layer function with all segments having slopes in $[-1, 1]$. We call this modified process the *truncated refinement process*, and the tree it produces, the *truncated tree*. Triangles (and corresponding tree nodes) that are no longer produced are said to have been *pruned*. We must show that the truncated refinement process still produces a plus discrepancy $\Theta(\sqrt{n} \log^{3/4} n)$.

Note that whenever we prune a node we must also prune all descendants of that node, so it is sufficient to bound the probability that leaves of the full tree are pruned; a leaf is pruned if and only if some ancestor (including the leaf itself) has a base slope outside $[-1, 1]$. Since the base slope of each node is either equal to that of its parent or is produced by adding $\pm s$ to that of its parent, the use of a random walk is suggested. A problem arises however. Although all of the central quadrilaterals we consider are disjoint and, therefore, the values of $\Delta^+(Q)$ for all such quadrilaterals are independent, the successive increments to the base slopes on a given path in the tree are not independent. To see this, consider the table below, which shows how the base slope of a child differs from that of its parent in various cases. The symbol "$\triangle$" indicates that the orientation of a triangle is such that it is above its base; "$\triangledown$" indicates that it is below its base.

Orientation of ABC	Left child		Right child	
	Triangle	Slope change	Triangle	Slope change
$\triangledown$	ADB ($\triangle$)	$-s$	BEC ($\triangle$)	$+s$
	AGF ($\triangledown$)	0	FHC ($\triangledown$)	0
$\triangle$	ADB ($\triangledown$)	$+s$	BEC ($\triangledown$)	$-s$
	AGF ($\triangle$)	0	FHC ($\triangle$)	0

A simple randomization, however, does produce the desired independence. Suppose we generate a random path down the full refinement tree by starting at the root and then repeatedly move to a child of the current node, using a coin flip to determine whether to move left or right. Consider the sequence of slope changes along this path. Since at each step we are equally likely to go to the left or the right and since we are equally likely to have chosen either orientation for the child during the refinement process, we see from the above table that, regardless of the orientation of the parent, at each step the change in slope is $-s$, 0, or $+s$, with probabilities $1/4$, $1/2$, $1/4$, respectively.

Hence let Z_j, $1 \leq j \leq k$, be i.i.d. random variables taking the values $-s$, 0, and $+s$ with probabilities $1/4$, $1/2$, and $1/4$, respectively. Then the probability that we reach a leaf that was not pruned is the probability that all of the partial sums $\sum_{j=1}^{i} Z_j$, $0 \leq i \leq k$, lie in $[-1,1]$. By using Theorem 2.7 (page 20), it is easy to see that this probability is bounded below by a positive constant. Hence, on average, a percentage $\Omega(1)$ of the leaves remain in the truncated refinement tree; in fact, since the presence of a leaf guarantees the presence of all of its ancestors, we can conclude that, on average, at least a percentage $\Omega(1)$ of the nodes at each level remains. Finally, it is not hard to see that the choice of whether a node, with central quadrilateral Q, is pruned is independent of the value of $\Delta(Q)$. Thus the truncation changes the expected plus discrepancy of the original full refinement process by only a factor of $\Theta(1)$. ∎

3.2.2 The upper bound

The following lemma applies Bernstein's bound (Theorem 2.6, page 19) and furnishes a basis for the probability estimates needed to prove the upper bound in (3.14). The notation Δf refers to the discrepancy of the lower layer defined by f.

Lemma 3.5 *Let f_1 and f_2 be two lower-layer functions with $\int_0^1 |f_1(x) - f_2(x)| dx = \alpha$. Then there exists a $c_0 > 0$ such that*

$$\Pr\{|\Delta f_1 - \Delta f_2| > z\} = \begin{cases} O(e^{-c_0 z^2 / \alpha n}) & \text{if } z \leq \alpha n, \\ O(e^{-c_0 z}) & \text{if } z > \alpha n. \end{cases}$$

Proof. Enumerate the points in S, and let R denote the region bounded entirely by f_1 and f_2 and having area α. Figure 3.10 illustrates the definition. Define $p_k = 0$ if the kth point of S is not in R; otherwise, $p_k = +1$ or -1 according as the kth point is a plus or minus. Then $\Delta(R) = \sum_{k=1}^{n} p_k$ is a sum of i.i.d. random variables on $[-1, +1]$, each having mean 0 and variance α. It is easily seen that, by symmetry, $|\Delta(R)|$ is equal in distribution to $|\Delta^+ f_1 - \Delta^+ f_2|$. Since $|\Delta f_1 - \Delta f_2| \leq |\Delta^+ f_1 - \Delta^+ f_2|$, the result follows easily from Theorem 2.6 (with $x\sqrt{n} = z$) applied to $\Delta(R)$:

$$\Pr\{|\Delta f_1 - \Delta f_2| > z\} \leq \Pr\{\Delta(R) > z\} \leq \exp\left(-\frac{z^2/2}{n\alpha + z/3}\right). \quad \blacksquare$$

The next lemma uses a convention that applies throughout the remainder of the proof: when we write "$g_1(n) = O(g_2(n))$ with high probability"

3.2. PROOF OF THE UP-RIGHT MATCHING ESTIMATE

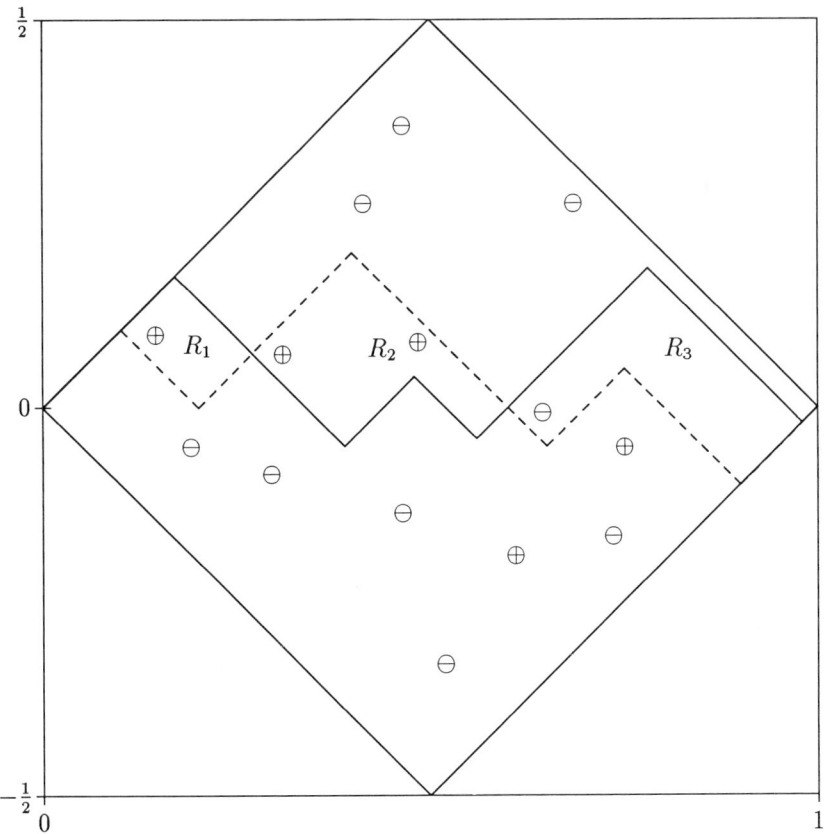

Figure 3.10: Illustration of the region R. R is $R_1 \cup R_2 \cup R_3$, and α is the area of R.

for given functions $g_1(n)$, $g_2(n)$, $n = 1, 2, \ldots$, we mean that there exist constants $\beta > 0$ and $c \geq 1$ such that for all n sufficiently large, $\Pr\{g_1(n) > \beta g_2(n)\} \leq 1/n^c$. Occasionally, we write *whp* as an abbreviation for "with high probability." Also, the symbol c will be used generically to denote constants; unless noted otherwise, constraints on c are determined by the immediate context only.

Lemma 3.6 *The following statement holds with high probability. Let $f_1 \in \mathcal{F}$, and let f_2 be any other function over $[0,1]$ such that for some $c > 0$, $|f_1(x) - f_2(x)| \leq c\sqrt{\lg n/n}$ uniformly in x, $0 \leq x \leq 1$. Then $|\Delta f_1 - \Delta f_2| = O(\sqrt{n \log n})$.*

Proof. Place a grid of squares of sizes $\sqrt{\lg n/n} \times \sqrt{\lg n/n}$ over the unit square, as shown in Figure 3.11. The number of points within a grid square entirely inside the rotated square is binomially distributed with mean $2 \lg n$. We find from standard[5] estimates for this distribution that *all* squares in the grid have $O(\log n)$ points with high probability.

Now in any column of the grid, f_1 intersects at most two squares (since $|f_1'(x)| = 1$ on the pieces of f_1), so $|f_1(x) - f_2(x)| \leq c\sqrt{\lg n/n}$ shows that the difference in the discrepancies of f_1 and f_2 within any column is concentrated in at most a constant number of squares. Then over $[0,1]$ the difference in the discrepancies is concentrated in at most $O(\sqrt{n/\log n})$ squares. The lemma follows at once from the fact that all squares have $O(\log n)$ points with high probability. ∎

Consider the grid introduced in the proof of Lemma 3.6. It is clear that for any function $f_1 \in \mathcal{F}$ we can construct another function $f_2 \in \mathcal{F}$ such that the vertices of f_2 coincide with vertices in the grid and such that $|f_1(x) - f_2(x)| \leq \sqrt{\lg n/n}$, for $0 \leq x \leq 1$. Let $\mathcal{F}^*$ be the subset of such functions. Figure 3.11 illustrates the construction. Clearly, by Lemma 3.6, (3.14) will be proved if we can show that

$$\mathsf{E}\left[\sup_{f \in \mathcal{F}^*} \Delta f\right] = O(\sqrt{n} \log^{3/4} n). \quad (3.15)$$

Elementary Fourier analysis shows that a lower-layer function $f \in \mathcal{F}$ may be represented by the sine series

$$f(x) = \sum_{i \geq 1} a_i \sin \pi i x \quad \text{for } 0 \leq x \leq 1. \quad (3.16)$$

---

[5] In particular, we could use Bernstein's bound Theorem 2.6 by treating the number of points as the sum of n i.i.d. 0-1 random variables X_i, i.e., X_i is 1 or 0 according as the ith point is or is not in a given grid square.

3.2. PROOF OF THE UP-RIGHT MATCHING ESTIMATE 67

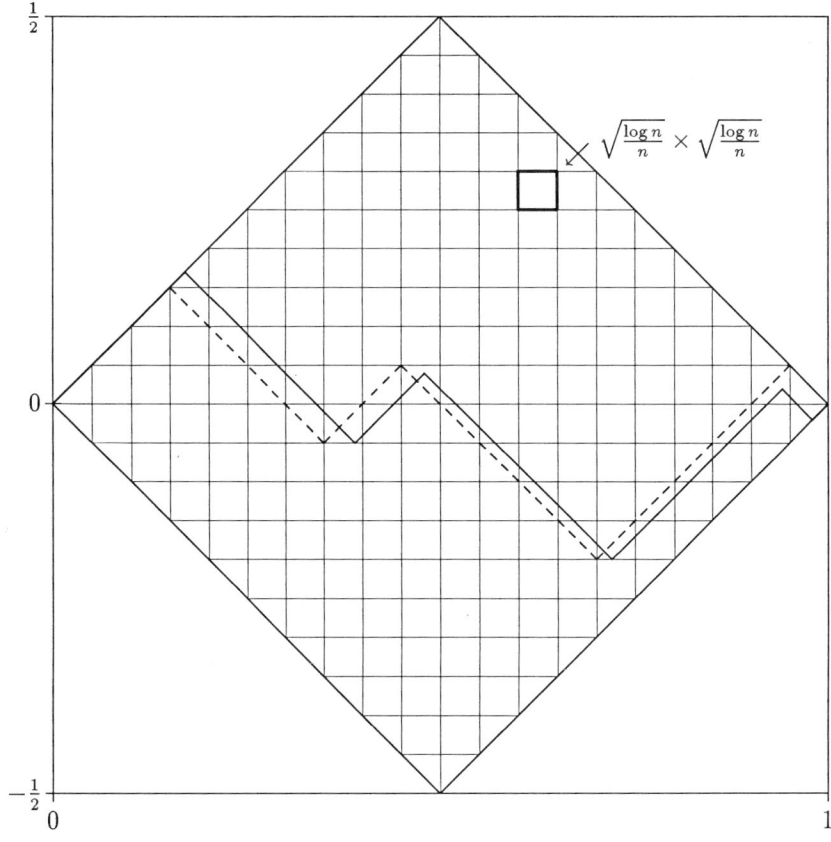

Figure 3.11: Approximating a function in $\mathcal{F}$ by a function in $\mathcal{F}^*$. The function shown as a solid line is in $\mathcal{F}$, and the approximating function shown as a dashed line is in $\mathcal{F}^*$.

These expansions play a key role in proving (3.14). Note that, since $|f'(x)| = 1$ on the pieces of f, we have

$$1 = \int_0^1 [f'(x)]^2 dx = \frac{\pi^2}{2} \sum_{i \geq 1} i^2 a_i^2,$$

so

$$\sum_{i \geq 1} i^2 a_i^2 = \frac{2}{\pi^2}. \tag{3.17}$$

Our final preliminary result describes the convergence in (3.16) for $f \in \mathcal{F}^*$, and is an immediate consequence of a result in [Jack41, page 21]. Let $f_n(x)$, $n \geq 1$, denote the nth partial sum in (3.16).

Lemma 3.7 *There exists a universal constant $c > 0$ such that for all $f \in \mathcal{F}^*$ and for all $x \in [0,1]$*

$$|f(x) - f_n(x)| = \frac{c}{\sqrt{n \log n}}.$$

With these preliminary results in hand, we are now ready to prove the theorem. As an application of the bound-that-usually-holds technique, we can write for any $c > 0$

$$\mathsf{E}[U_n] \leq c\sqrt{n}\lg^{3/4} n + n\Pr\{U_n > c\sqrt{n}\lg^{3/4} n\},$$

since trivially $U_n \leq n$.

Then, since U_n is stochastically bounded by $\sup_{L \in \mathcal{L}} \Delta(L)$, it is enough to prove that $\sup_{L \in \mathcal{L}} \Delta(L) = O(\sqrt{n} \log^{3/4} n)$ with high probability. This in turn will be proved if we can show that $\sup_{f \in \mathcal{F}^*} \Delta f = O(\sqrt{n} \log^{3/4} n)$ with high probability (see (3.15)). This last result is proved below.

Let $f^{(1)}, f^{(2)}, \ldots, f^{(\lfloor \lg n \rfloor)}$ be successively better approximations of $f \in \mathcal{F}^*$ defined by

$$f^{(k)}(x) = \sum_{i=1}^{2^{k+1}} a_i(k) \sin \pi i x \quad \text{for } 1 \leq k \leq \lfloor \lg n \rfloor, \tag{3.18}$$

where $a_i(k)$ is a_i truncated to the $\lfloor 3k/2 + \frac{1}{2} \lg \lg n \rfloor$ most significant bits in the fractional part of its binary representation. (Hereafter, we shall omit the floor notation and treat the affected quantities as integers; extension of the analysis to noninteger values is trivial and influences only hidden constants.) By (3.18) differences in accuracy are bounded by

$$|a_i(k) - a_i(k-1)| \leq 2^{-3(k-1)/2 + (\lg \lg n)/2} = \frac{1}{2^{3(k-1)/2}\sqrt{\lg n}}. \tag{3.19}$$

3.2. PROOF OF THE UP-RIGHT MATCHING ESTIMATE

Clearly, for given f, the possible functions $f^{(k)}$ make up a finite set. We will be counting certain subsets of these functions in terms of properties defined by

$$t(f^{(k)}) = \sum_{j=0}^{k} r_j(f^{(k)}) 2^{j-k} \text{ with } r_j(f^{(k)}) = \begin{cases} a_1^2 + a_2^2 & \text{for } j = 0, \\ 4^j \sum_{i=2^j+1}^{2^{j+1}} a_i^2(k) & \text{for } j > 0. \end{cases} \quad (3.20)$$

From (3.17), we have

$$\sum_{j=0}^{\lg n} r_j(f^{(k)}) = a_1^2(k) + \sum_{j=0}^{\lg n} \sum_{i=2^j+1}^{2^{j+1}} [2^j a_i(k)]^2 \le \sum_{i=1}^{\infty} i^2 a_i^2 = \frac{2}{\pi^2}, \quad (3.21)$$

and hence

$$\sum_{k=1}^{\lg n} t(f^{(k)}) = \sum_{k=1}^{\lg n} \sum_{j=0}^{k} r_j(f^{(k)}) 2^{j-k} \le \sum_{j=0}^{\lg n} r_j(f^{(\lg n)}) \sum_{k=j}^{\lg n} 2^{j-k} \le \frac{4}{\pi^2}. \quad (3.22)$$

In particular, note that (3.21) implies, for all i, that $|a_i| < 1$.

The following lemma comprises the combinatorial part of the proof.

Lemma 3.8 *There exist a universal constant c and a mapping from $\mathcal{F}^*$ into $\mathcal{R}^{\lg n}$, with values denoted by $\big(s_1(f), s_2(f), \ldots, s_{\lg n}(f)\big)$, such that for each $f \in \mathcal{F}^*$, we have $\sum_{k=1}^{\lg n} s_k(f) \le c$,*

$$\begin{aligned} \int_0^1 |f^{(1)}(x)| dx &\le \sqrt{s_1(f)}/2, \\ \int_0^1 |f^{(k)}(x) - f^{(k-1)}(x)| dx &\le \sqrt{s_k(f)}/2^k \quad \text{for } 2 \le k \le \lg n, \end{aligned} \quad (3.23)$$

and such that if $\eta_k(\sigma)$ denotes the number of functions $g^{(k)}$, for which $g \in \mathcal{F}^*$, such that $s_k(g) \le \sigma$, then

$$\eta_k(\sigma) \le \min \left\{ (\sigma \lg n)^{2^k}, 2^{\sqrt{n/\lg n}} \right\} \quad \text{for } 1 \le k \le \lg n. \quad (3.24)$$

Proof. We will show that, for a large enough constant $\gamma > 0$, the mapping

$$s_k(f) = \gamma(t(f^{(k)}) + 1/\lg n) \quad (3.25)$$

has the properties claimed in the lemma. First, note that $\sum_{k=1}^{\lg n} s_k(f) = O(1)$ by use of (3.22).

To prove (3.23), consider a function $f \in \mathcal{F}^*$ and write from (3.18), for $2 \leq k \leq \lg n$,

$$f^{(k)}(x) - f^{(k-1)}(x) = \sum_{i=2^k+1}^{2^{k+1}} a_i(k) \sin \pi i x + \sum_{i=1}^{2^k} [a_i(k) - a_i(k-1)] \sin \pi i x. \quad (3.26)$$

Let $g_k(x)$ and $h_k(x)$ denote the first and second sums in (3.26), respectively. We have from (3.20), $\int_0^1 g_k^2(x) dx = \frac{1}{2} \sum_{i=2^k+1}^{2^{k+1}} a_i^2(k) = \frac{1}{2} 4^{-k} r_k(f^{(k)})$, so by Schwarz's inequality, with $r_k \equiv r_k(f^{(k)})$,

$$\int_0^1 |g_k(x)| dx \leq 2^{-k} \sqrt{r_k}. \quad (3.27)$$

Next, by (3.19),

$$\int_0^1 h_k^2(x) dx \leq \frac{1}{2} \sum_{i=1}^{2^k} [a_i(k) - a_i(k-1)]^2 \leq \frac{2^{k+2}}{2^{3k} \lg n},$$

so

$$\int_0^1 |h_k(x)| dx \leq 2^{-k} \sqrt{4/\lg n}. \quad (3.28)$$

Now add (3.27) and (3.28) and note that $\sqrt{r_k} + \sqrt{4/\lg n} \leq \sqrt{2(r_k + 4/\lg n)}$ by Cauchy's inequality. Thus (3.26)–(3.28) yield

$$\int_0^1 |f^{(k)}(x) - f^{(k-1)}(x)| dx \leq \sqrt{2(r_k + 4/\lg n)}. \quad (3.29)$$

But since for any $\gamma > 4$, (3.25) and (3.20) imply

$$s_k(f) = \gamma \left(t(f^{(k)}) + 1/\lg n \right) \geq 2(r_k + 4/\lg n),$$

(3.29) yields the second equation in (3.23). (We omit the proof of the first, which is similar but easier.)

For the second part of the lemma, we note, first, that, for all k, $\eta_k(\sigma) \leq 2^{\sqrt{n/\lg n}}$ follows from the definition of $\mathcal{F}^*$—there are at most $2^{\sqrt{n/\lg n}}$ functions $f \in \mathcal{F}^*$, since the vertices of these functions are restricted to the vertices of a $\sqrt{n/\lg n} \times \sqrt{n/\lg n}$ grid on the unit square. To prove $\eta_k(\sigma) \leq (\sigma \lg n)^{2^k}$, we first establish a bound on $\mu_k(\tau)$, defined as the number of functions $g^{(k)}$, $g \in \mathcal{F}^*$, such that $t(g^{(k)}) \leq \tau$; our bound on $\mu_k(\tau)$ will only be required to be valid when $\tau \geq 1/\lg n$.

Consider a function $f \in \mathcal{F}^*$ and rewrite $f^{(k)}$ in (3.18) as

$$f^{(k)}(x) = a_1(k) \sin \pi x + a_2(k) \sin 2\pi x + \sum_{j=1}^{k} \sum_{i=2^j+1}^{2^{j+1}} a_i(k) \sin \pi i x. \quad (3.30)$$

3.2. PROOF OF THE UP-RIGHT MATCHING ESTIMATE

Now consider the number of possibilities in the jth inner sum, i.e., the number of vectors of coefficients $a_i(k)$, $2^j + 1 \leq i \leq 2^{j+1}$, $1 \leq j \leq k$. It is more convenient to work with the numbers $b_i(k) = a_i(k)2^{3k/2}\sqrt{\lg n}$, since these are integers by definition of the $a_i(k)$. By (3.20), the $b_i(k)$ satisfy

$$b_1^2(k) + b_2^2(k) = 2^{3k} \lg n \left(a_1^2(k) + a_2^2(k)\right) \leq 2^{3k} r_0(f^{(k)}) \lg n,$$

$$\sum_{i=2^j+1}^{2^{j+1}} b_i^2(k) = 2^{3k} \lg n \sum_{i=2^j+1}^{2^{j+1}} a_i^2(k) \leq 2^{3k-2j} r_j(f^{(k)}) \lg n$$

for $1 \leq k \leq \lg n$. (3.31)

Now divide through by 2^{4k-3j} and sum over $j \geq 0$. Assuming that $t(f^{(k)}) \leq \tau$, this leads to

$$\sum_{j=1}^{k} 2^{3j-4k} \sum_{i=2^j+1}^{2^{j+1}} b_i^2(k) \leq \lg n \sum_{j=1}^{k} r_j(f^{(k)})2^{j-k} = t(f^{(k)}) \lg n \leq \tau \lg n. \quad (3.32)$$

Thus the number of functions $f^{(k)}$ with $t(f^{(k)}) \leq \tau$ is bounded by the number of pairs $\left(a_1(k), a_2(k)\right)$ times the number of vectors $\left(b_3(k), \ldots, b_{2^{k+1}}(k)\right)$ satisfying (3.32). The former quantity is $2^{3k} \lg n$, as can be seen from the definition of $a_i(k)$. The latter quantity is the number of lattice points in a $(2^{k+1} - 2)$-dimensional ellipsoid with 2^j axes of lengths $2\sqrt{\tau \lg n 2^{4k-3j}}$, for each $j = 1, 2, \ldots, k$, which, in turn, is approximately the volume of the ellipsoid. Now a d-dimensional ellipsoid with axis lengths $l_1, \ldots, l_d$ has volume

$$\left(\prod_{j=1}^{d} \frac{l_i}{2}\right) \frac{\pi^{d/2}}{(d/2)!} \leq \left(\prod_{j=1}^{d} \frac{l_i}{2}\right) \left(\frac{2\pi e}{d}\right)^{d/2},$$

where the inequality is obtained from Stirling's formula. With $d = 2^{k+1} - 2$ and the l_i given above, this yields

$$\mu_k(\tau) \approx 2^{3k} \lg n \left(\frac{2\pi e}{d}\right)^{d/2} \prod_{i=1}^{d} \frac{l_i}{2}.$$

It is easy to verify that, if $\tau \geq 1/\lg n$, replacing the constant $2\pi e$ by a suitably larger constant allows us to drop the $2^{3k} \lg n$ factor and convert the volume approximation to an upper bound. Substituting for d and the l_i, we then find that for some constant γ,

$$\mu_k(\tau) \leq \frac{(\gamma \tau \lg n)^{2^k-1}}{(2^{k+1}-2)^{2^k-1}} \prod_{j=1}^{k} 2^{(4k-3j)2^j/2} \quad \text{for } \tau \geq 1/\lg n.$$

Routine algebra then shows that there is a constant γ such that

$$\mu_k(\tau) \leq (\gamma\tau \lg n)^{2^k} \quad \text{for } \tau \geq 1/\lg n. \tag{3.33}$$

By (3.25) we know that $\eta_k(\sigma)$, the number of $f \in \mathcal{F}^*$ satisfying $s_k(f) \leq \gamma$, is the number of $f \in \mathcal{F}^*$ satisfying $t(f^{(k)}) \leq \sigma/\gamma - 1/\lg n$, which is

$$\mu_k(\sigma/\gamma - 1/\lg n).$$

In the case that $\sigma/\gamma < 1/\lg n$, this value is clearly 0, so (3.24) holds trivially. If $\sigma/\gamma \geq 1/\lg n$, using (3.33), we have $\mu_k(\frac{\sigma}{\gamma} - \frac{1}{\lg n}) \leq \mu_k(\frac{\sigma}{\gamma}) \leq (\sigma \lg n)^{2^k}$, and again (3.24) holds. ∎

As a trivial extension of Lemma 3.8, it is convenient to assume for each f that $s_k(f)$ is a positive multiple of $1/\lg n$, $1 \leq k \leq \lg n$.

We turn now to the probabilistic part of the proof. Consider any function $f^{(\lg n)}$ with $f \in \mathcal{F}^*$. Comparing $f^{(\lg n)}$ and the partial sums f_n, we have by the definition of the $a_i(k)$ that $|a_i - a_i(\lg n)| = O\left(1/(\sqrt{\log n}\, n^{3/2})\right)$ and hence

$$|f_n(x) - f^{(\lg n)}(x)| \leq \sum_{i=1}^{n}|a_i - a_i(\lg n)| = O\left(\frac{1}{\sqrt{n\log n}}\right) \quad \text{for } 0 \leq x \leq 1. \tag{3.34}$$

By Lemma 3.7 we have $|f(x) - f_n(x)| = O(1/\sqrt{n\log n})$, $0 \leq x \leq 1$. This together with (3.34) yields $|f(x) - f^{(\lg n)}(x)| = O(1/\sqrt{n\log n})$, $0 \leq x \leq 1$. We conclude from Lemma 3.6 applied to $f^{(\lg n)}$ and f that if

$$\sup_{f \in \mathcal{F}^*} \Delta f^{(\lg n)} = O(\sqrt{n}\log^{3/4} n) \text{ whp}, \tag{3.35}$$

then $\sup_{f \in \mathcal{F}^*} \Delta f = O(\sqrt{n}\log^{3/4} n)$ whp as well. We prove (3.35) below.

Consider any $f \in \mathcal{F}^*$ and write

$$\Delta f^{(\lg n)} = \sum_{k=1}^{\lg n}(\Delta f^{(k)} - \Delta f^{(k-1)}), \tag{3.36}$$

with $\Delta f^{(0)} \equiv 0$. Below, we introduce numbers $q_k = q_k(s_k(f))$, $1 \leq k \leq \lg n$, such that $\sum_{k=1}^{\lg n} s_k(f) \leq c$ implies $\sum_{k=1}^{\lg n} q_k = O(\sqrt{n}\log^{3/4} n)$ for all $f \in \mathcal{F}^*$. If f is such that $\Delta f^{(\lg n)} > \sum_{k=1}^{\lg n} q_k$, then there exist k and σ, with $1 \leq k \leq \lg n$, $1/\lg n \leq \sigma \leq c$ (with c as given in Lemma 3.8), such that $s_k(f) = \sigma$ and $\Delta f^{(k)} - \Delta f^{(k-1)} > q_k(\sigma)$. Over all $f \in \mathcal{F}^*$, the number of pairs of functions $(f^{(k)}, f^{(k-1)})$ for given k, σ, and $s_k(f) = \sigma$ is at most $\eta_k(\sigma)$, so by Boole's

3.2. PROOF OF THE UP-RIGHT MATCHING ESTIMATE

inequality,

$$\Pr\left\{\max_{f\in\mathcal{F}^*}\Delta f^{(\lg n)} > \sum_{k=1}^{\lg n} q_k\bigl(s_k(f)\bigr)\right\}$$
$$\leq c\lg^2 n \max_{\substack{1\leq k\leq \lg n \\ 1/\lg n \leq \sigma \leq c}} \max_{\{f\in\mathcal{F}^*:\ s_k(f)=\sigma\}} \eta_k(\sigma)\Pr\{\Delta f^k - \Delta f^{(k-1)} > q_k(\sigma)\}, \tag{3.37}$$

where the $c\lg^2 n$ factor comes from the $\lg n$ values of k and the at most $c\lg n$ values of σ (recall that the $s_k(f)$ are chosen as multiples of $1/\lg n$). With $\eta_k(\sigma)$ bounded by (3.24), and with $q_k(\sigma)$ as defined below, we will show that

$$\min\left\{2\sqrt{n/\lg n}, (\sigma\lg n)^{2^k}\right\}\Pr\{\Delta f^{(k)} - \Delta f^{(k-1)} > q_k(\sigma)\} \leq 1/n^2 \tag{3.38}$$

for any choice of k, σ, and $f \in \mathcal{F}^*$ with $s_k(f) = \sigma$. Since $\lg^2 n/n^2 = O(1/n)$, the proof of (3.35) will be complete once we have verified that, in the left-hand side of (3.37), $\sum_{k=1}^{\lg n} q_k(s_k(f)) = O(\sqrt{n}\log^{3/4} n)$ for all $f \in \mathcal{F}^*$.

Now consider the pair of functions $(f^{(k)}, f^{(k-1)})$ for an f such that $s_k(f) = \sigma$. To apply Lemma 3.5 to (3.38), let $f^{(k)}$ and $f^{(k-1)}$ be f_1 and f_2, respectively, and let $\alpha = \sqrt{\sigma}/2^k$ from (3.23). Define $u_k = u_k(\sigma)$ by

$$u_k^2 \equiv c_0^{-1}n\sqrt{\sigma}\left(\ln(\sigma\lg n) + 2^{-k+1}\ln n\right), \tag{3.39}$$

where c_0 is as in Lemma 3.5. We consider two cases. If $u_k \leq \alpha n$, substitute (3.39) into the first of the bounds in Lemma 3.5 and then substitute the result into the left-hand side of (3.38). A little algebra shows that

$$\min\left\{2\sqrt{n/\lg n}, (\sigma\lg n)^{2^k}\right\}\Pr\{\Delta f^{(k)} - \Delta f^{(k-1)} > u_k\}$$
$$= O\left((\sigma\lg n)^{2^k}e^{-c_0 u_k^2 2^k/n\sqrt{\sigma}}\right) = O(1/n^2), \tag{3.40}$$

as desired. To take care of the other case, $u_k > \alpha n$, we put $q_k = q_k(\sigma) = u_k + v$, where

$$v = c_0^{-1}\left(\sqrt{n/\lg n} + 2\lg n\right)\ln 2. \tag{3.41}$$

For, if $q_k > \alpha n$, then Lemma 3.5 and substitution into the left-hand side of (3.38) give

$$\min\left\{2\sqrt{n/\lg n}, (\sigma\lg n)^{2^k}\right\}\Pr\{\Delta f^{(k)} - \Delta f^{(k-1)} > q_k\} = O\left(2\sqrt{n/\lg n}e^{-c_0 v}\right)$$
$$= O(1/n^2),$$

again as desired. Thus (3.38) is established.

It remains to show that $\sum_{k=1}^{\lg n} q_k = \sum_{k=1}^{\lg n}(u_k + v) = O(\sqrt{n}\log^{3/4} n)$ for all $f \in \mathcal{F}^*$. Recall that $\sigma = s_k(f)$. For the contribution of the u_k use Cauchy's inequality, abbreviate $s_k = s_k(f)$, and write

$$u_k = O\left(\sqrt{n}s_k^{1/4}\sqrt{\log(s_k \log n)} + \sqrt{n}s_k^{1/4}2^{-k/2}\sqrt{\log n}\right). \tag{3.42}$$

Since s_k is bounded by a constant, the contribution to $\sum u_k$ of the second term in (3.42) is easily seen to be $O(\sqrt{n \log n})$. By Lemma 3.8 the sum of the s_k, $1 \leq k \leq \lg n$, is at most a constant c, so the contribution of the first term is $O(w_n)$, where

$$w_n = \max_{\{z_k\}} \left\{ \sqrt{n} \sum_{k=1}^{\lg n} z_k^{1/4} \lg(z_k \lg n) : z_k \geq \frac{1}{\lg n} \text{ and } \sum_{k=1}^{\lg n} z_k \leq c \right\}. \tag{3.43}$$

A calculation shows that the function $w(z) = z^{1/4}\lg(z \lg n)$ is increasing and concave (i.e., $w''(z) \leq 0$) for all $z \geq 1/\lg n$. Then by Jensen's inequality the maximum in (3.43) is achieved by putting all z_k equal to $c/\lg n$, so

$$w_n = \sqrt{n} \sum_{k=1}^{\lg n} \frac{c^{1/4} \lg c}{\lg^{1/4} n} = O(\sqrt{n}\log^{3/4} n),$$

and hence $\sum_{k=1}^{\lg n} u_k = O(\sqrt{n}\log^{3/4} n)$. It can easily be seen by inspection that $v \lg n = O(\sqrt{n \log n})$, so (3.35), and hence the upper bound, is proved. ∎

Chapter 4
Scheduling and Partitioning

4.1 Analysis of classical greedy heuristics

In the previous chapter we studied the simplest scheduling heuristic, LS, in a number of examples. The results confirmed that for fixed m the expected relative error of LS tends to 0 as $n \to \infty$. However, for the *absolute error*

$$A^{\text{LS}}(L_n, m) = \text{LS}(L_n, m) - \text{OPT}(L_n, m), \qquad (4.1)$$

this is not the case. Several more difficult probabilistic analyses have concerned better heuristics, H, for which $A^H(L_n, m) \to 0$ almost surely as $n \to \infty$ with m fixed. Such a heuristic is the LPT rule described in Section 1.4.1, where the tasks X_i are assigned to processors in decreasing order of size, i.e., the tasks are assigned in the order $X_{(n)} \geq X_{(n-1)} \geq \cdots \geq X_{(1)}$. Intuitively, the smaller tasks assigned later by LPT can be used to patch up the differences between the workloads of the processors. One expects LPT to do much better than LS, and, indeed, as shown below, $A^{\text{LPT}}(L_n, m) \to 0$ as $n \to \infty$ under very mild conditions on the distribution F of the X_i. (Since we will often consider the almost-sure convergence of sequences of random variables, we will use the standard abbreviation (a.s.) to mean "almost surely.")

The bounds of (2.38) and (2.39) are not strong enough to lead to the desired results. A number of authors, such as Loulou [Loul84a], have found that tighter bounds can be obtained by looking directly at the idle time of the processors; although various notations have been used, here we cast all of these results in terms of the *average idle time* $D^{\text{LPT}}(L_n, m)$, which is the average over all processors of the difference between this processor's workload and the makespan; more precisely, we define $D^{\text{LPT}}(L_n, m)$ by

$$\text{LPT}(L_n, m) = \frac{1}{m} \sum_{i=1}^{n} X_i + D^{\text{LPT}}(L_n, m), \qquad (4.2)$$

so from (2.38) we have

$$A^{\text{LPT}}(L_n, m) \leq D^{\text{LPT}}(L_n, m). \tag{4.3}$$

The result below verifies that LPT is asymptotically optimal in the strong absolute-error sense. Frenk and Rinnooy Kan [FR87, RF86] noted, as did Loulou [Loul84a], that

$$D^{\text{LPT}}(L_n, m) \leq \max_{1 \leq i \leq n} \left\{ X_{(i)} - \frac{1}{m} \sum_{k=1}^{i} X_{(k)} \right\}. \tag{4.4}$$

To see this, let i be the largest index such that $X_{(i)}$ runs until the end of the schedule. Then, just after $X_{(i)}$ is scheduled, the average idle time is at most $(m-1)X_{(i)}/m$. Each task $X_{(k)}$ scheduled after $X_{(i)}$ reduces the average idle time by $X_{(k)}/m$. The bound in (4.4) provides a powerful tool. For instance, let us assume that F is strictly increasing in $(0, \delta)$ for some $\delta > 0$, and $\mathsf{E}[X_i] < \infty$. Bounding the right-hand side of (4.4) by

$$X_{(\lfloor \epsilon n \rfloor)} + \max \left\{ X_{(n)} - \frac{1}{m} \sum_{k=1}^{\lfloor \epsilon n \rfloor} X_{(k)}, 0 \right\}, \quad 0 < F^{-1}(\epsilon) < \delta, \tag{4.5}$$

we observe that the first term in (4.5) converges (a.s.) to $F^{-1}(\epsilon)$ as $n \to \infty$, and can be made arbitrarily small by an appropriate choice of ϵ (see [Serf80], for example). Also, $X_{(n)}/n \to 0$ (a.s.) (since $\mathsf{E}[X_i] < \infty$) and $\sum_{k=1}^{\lfloor \epsilon n \rfloor} X_{(k)}/n$ converges (a.s.) to a positive constant as $n \to \infty$ for every $\epsilon > 0$. Thus the first term within the maximization in (4.5) tends to $-\infty$ (a.s.), and we have

Theorem 4.1 ([FR87]) *If $\mathsf{E}[X_i] < \infty$ and, for some $\delta > 0$, $F(x)$ is strictly increasing in $(0, \delta)$, then $A^{\text{LPT}}(L_n, m) \to 0$ (a.s.) as $n \to \infty$.*

We turn next to the analysis of the more interesting *rates* of convergence. The first such result below focuses on the expected value when the X_i are i.i.d. from $U(0, 1)$. The analysis exploits the relation discussed in Section 2.7.1 between the uniform distribution and the Poisson process.

We need a preliminary combinatorial result.

Lemma 4.2 *In the event that the X_i satisfy $\alpha i < x_i < \beta i$ for $m \leq i \leq n$, where $0 < \alpha < \beta$, we have $D^{\text{LPT}}(L_n, m) \leq m\beta^2/\alpha$.*

Proof. Let

$$D_i = X_{(i)} - \frac{1}{m} \sum_{k=1}^{i} X_{(k)},$$

4.1. ANALYSIS OF CLASSICAL GREEDY HEURISTICS

so that
$$D^{\text{LPT}}(L_n, m) \leq \max_{1 \leq i \leq n} D_i,$$
by (4.4). Also define
$$d_i = \begin{cases} \beta m & \text{for } i < m, \\ \beta i - \dfrac{1}{m} \sum_{m \leq k \leq i} \alpha k & \text{for } i \geq m, \end{cases} \quad (4.6)$$

so that in the specified event we have $D_i \leq d_i$, and hence
$$D^{\text{LPT}}(L_n, m) \leq \max_{1 \leq i \leq n} d_i. \quad (4.7)$$

Now $d_i - d_{i-1} = \beta - \frac{1}{m}\alpha i$ for $i > m$, so it is not hard to see that a value of i that maximizes d_i is the largest i for which $\beta - \frac{1}{m}\alpha i \geq 0$, i.e., $i = \lfloor \beta m/\alpha \rfloor$. (Note that this is at least m, since $\beta > \alpha$.) Now by (4.6) we have $d_{\lfloor \beta m/\alpha \rfloor} \leq \beta \lfloor \beta m/\alpha \rfloor \leq m\beta^2/\alpha$, so (4.7) gives the lemma. ∎

Theorem 4.3 ([CFL84a]) *If $F(x)$ is $U(0,1)$, then $\mathsf{E}[D^{\text{LPT}}(L_n, m)] \leq c_m m/(n+1)$, where c_m is bounded and $\lim_{m \to \infty} c_m = 1$.*

Proof. Let $\hat{X}_1, \hat{X}_2, \ldots$ be the successive epochs of a Poisson process with parameter 1; i.e., the differences $\hat{X}_{i+1} - \hat{X}_i$, $i = 0, 1, 2, \ldots$, with $\hat{X}_0 = 0$, are i.i.d. random variables having an exponential distribution with mean 1. Given $\hat{X}_{(n+1)}$, the random variables $\hat{X}_i/\hat{X}_{n+1}$ are independent samples from $U(0,1)$, so we have
$$\mathsf{E}[\hat{D} \mid \hat{X}_{n+1} = x] = x\mathsf{E}[D^{\text{LPT}}(L_n, m)],$$
where $\hat{D}$ is the average idle time within the LPT schedule of n tasks whose order statistics are given by the $\hat{X}_i$, $1 \leq i \leq n$. Since $\mathsf{E}[\hat{X}_{n+1}] = n+1$, we have $\mathsf{E}[D^{\text{LPT}}(L_n, m)] = \mathsf{E}[\hat{D}]/(n+1)$. The remainder of the proof shows that $\mathsf{E}[\hat{D}] \leq c_m m$, where c_m is a bounded function of m that approaches 1 as $m \to \infty$.

From Chernoff's bound (Theorem 2.4, page 16) and (2.13) we remember that
$$\Pr\{\hat{X}_i \leq \alpha i\} \leq \bigl(u(\alpha)\bigr)^i, \quad (4.8)$$
and
$$\Pr\{\hat{X}_i \geq \beta i\} \leq \bigl(u(\beta)\bigr)^i, \quad (4.9)$$
where
$$u(x) = xe^{1-x}.$$

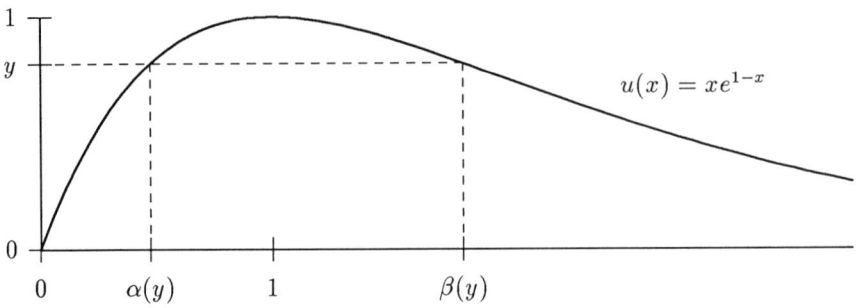

Figure 4.1: The function $u(x)$ and two inverse functions, α and β.

Note that as x increases, $u(x)$ increases monotonically from 0 to its maximum of $u(1) = 1$, and then decreases monotonically to 0 thereafter. (See Figure 4.1.) Thus, to each $u \in (0,1)$ there correspond two values of x, to be denoted $\alpha(u)$ and $\beta(u)$ (with $\alpha(u) < \beta(u)$), which are the functions inverse to $u(x)$; at $u = 1$ the two functions coincide, $\alpha(1) = \beta(1) = 1$. Hence if we let
$$\xi = \min_{i=m}^{n} u(\hat{X}_i/i),$$
the distribution G_m of ξ satisfies
$$\begin{aligned}
G_m(x) &= \Pr\{\xi \leq x\} \\
&= \Pr\left\{\exists i \in \{m, m+1, \ldots, n\},\ \hat{X}_i \leq \alpha(x)i \text{ or } \hat{X}_i \geq \beta(x)i\right\} \\
&\leq \sum_{i=m}^{n} \left(\Pr\{\hat{X}_i \leq \alpha(x)i\} + \Pr\{\hat{X}_i \geq \beta(x)i\}\right) \\
&\leq \sum_{i=m}^{n} \left(\bigl(u(\alpha(x))\bigr)^i + \bigl(u(\beta(x))\bigr)^i\right) \leq \sum_{i=m}^{n} 2x^i \leq \frac{2x^m}{1-x}.
\end{aligned}$$
Of course, we also have $G_m(x) \leq 1$, so
$$G_m(x) \leq \min\left\{\frac{2x^m}{1-x}, 1\right\}. \tag{4.10}$$
Then, applying Lemma 4.2, we have
$$\mathsf{E}[\hat{D}] \leq m c_m,$$
where
$$c_m = \int_0^1 \frac{\beta^2(x)}{\alpha(x)}\,dG_m(x) = \int_0^1 K(x)\,dG_m(x),$$

4.1. ANALYSIS OF CLASSICAL GREEDY HEURISTICS

and $K(x) = \beta^2(x)/\alpha(x)$. We now estimate this integral. Intuitively, for large m, the increase in G_m is concentrated near 1, where $K(x)$ is approximately 1, so we expect the integral to be approximately 1. Motivated by this intuition, we split the integral into two parts, one over $[0, x_m]$ and the other over $[x_m, 1]$, where $x_m = 1 - m^{-1/2}$. (This value of x_m is chosen so that $x_m \to 1$ as $m \to \infty$, but slowly enough so that most of the increase in G_m is to the right of x_m.)

It is not hard to see that $K(x)$ is continuous and decreasing over $(0, 1]$, and that $K(x) \sim e \ln^2 x / x$ for x near 0. Thus we have, for some constant C, $K(x) \leq C(1 + e \log^2 x / x)$. Hence

$$\int_0^{x_m} K(x) \, dG_m(x) \leq C \int_0^{x_m} \frac{\log^2 x}{x} \, dG_m(x) + C \int_0^{x_m} dG_m(x)$$

$$\leq C \int_0^{x_m} \frac{\log^2 x}{x} \, d\frac{2x^m}{1 - x_m} + CG_m(x_m)$$

$$\leq \frac{2C}{1 - x_m} \int_0^{x_m} \frac{\log^2 x}{x} m x^{m-1} \, dx + CG_m(x_m)$$

$$\leq \frac{2mC}{1 - x_m} \int_0^{x_m} x^{m-2} \log^2 x \, dx + CG_m(x_m)$$

$$\leq \frac{mC'}{1 - x_m} \frac{x_m^{m-1} \log^2 x_m}{(m-1)} + C' \frac{x_m^m}{1 - x_m}, \qquad (4.11)$$

where C' is some new constant. Since $x_m^m \sim e^{-\sqrt{m}}$, it is not difficult to see that this expression is bounded by a constant independent of m (assuming of course $m \geq 2$), and that it approaches 0 as $m \to \infty$.

Over the remaining interval, $[x_m, 1]$, we have

$$\int_{x_m}^1 K(x) \, dG_m(x) = K(\bar{x})\big(G_m(1) - G_m(x_m)\big) \qquad (4.12)$$

for some $\bar{x} \in [x_m, 1]$. For large m this means that $\bar{x}$ approaches 1, so since $K(x) \sim 1$ near 1 we know that $K(\bar{x}) \sim 1$. Also, for large m we have $G_m(1) - G_m(x_m) = 1 - G_m(x_m) \sim 1$, so the integral over $[x_m, 1]$ approaches 1.

Thus c_m, which is the integral over the entire interval $[0, 1]$, satisfies the conditions of the theorem. ∎

The next result, although it says less about multiplicative constants, generalizes Theorem 4.3 in two important ways. First, the result applies to any distribution of the form $F(x) = x^a$, $0 \leq x \leq 1$, for some $a > 0$, and, second, it gives information about the rate of convergence for *all* moments of $D^{\text{LPT}}(L_n, m)$. The result appears in the work of Boxma [Boxm84], but we use the proof in [RF86], where efficient use is made of (4.4).

Theorem 4.4 ([Boxm84, RF86]) *If* $F(x) = x^a$, $0 \le x \le 1$, *for some* $0 < a < \infty$, *then*
$$\mathsf{E}[A^{\text{LPT}}(L_n, m)^p] = O(n^{-p/a}).$$

Proof. We bound $\mathsf{E}[D^{\text{LPT}}(L_n, m)^p]$ using (4.4); then the theorem will follow from the fact that, by (4.3), $\mathsf{E}[A^{\text{LPT}}(L_n, m)^p] \le \mathsf{E}[D^{\text{LPT}}(L_n, m)^p]$.

Let $Z_{(k)}$ denote the kth-smallest order statistic of n independent samples from $U(0,1)$. Then, trivially, the $X_{(i)}$, $1 \le i \le n$, are equal in distribution to $Z_{(i)}^{1/a}$. Thus, taking the expected value of (4.4), we have

$$\mathsf{E}[D^{\text{LPT}}(L_n, m)^p] \le d_{n,m} \equiv \mathsf{E}\left[\left(\max_{1\le i\le n}\left\{Z_{(i)}^{1/a} - \frac{1}{m}\sum_{j=1}^{i} Z_{(j)}^{1/a}\right\}\right)^p\right]. \quad (4.13)$$

Let $\mathcal{E}_n$ be the event that the maximum in (4.13) is achieved at $i = n$. To develop a recurrence, we treat $\mathcal{E}_n$ separately; using the fact that the expression under the rightmost expectation is bounded by 1, we bound (4.13) by

$$d_{n,m} \le \Pr\{\mathcal{E}_n\} + \mathsf{E}\left[Z_{(n)}^{p/a}\left(\max_{1\le i\le n-1}\left\{\left(\frac{Z_{(i)}}{Z_{(n)}}\right)^{1/a} - \frac{1}{m}\sum_{j=1}^{i}\left(\frac{Z_{(j)}}{Z_{(n)}}\right)^{1/a}\right\}\right)^p\right]. \quad (4.14)$$

We have factored out $Z_{(n)}^{p/a}$ so as to facilitate the observation that, given $Z_{(n)}$, the ratios $Z_{(i)}/Z_{(n)}$, $1 \le i \le n-1$, are distributed as the order statistics of $n-1$ independent samples from $U(0,1)$. Then, since $\mathsf{E}[Z_{(n)}^{p/a}] = n/(n + p/a)$, we can rewrite (4.14) as

$$d_{n,m} \le \Pr\{\mathcal{E}_n\} + \frac{n}{n + p/a} d_{n-1,m}. \quad (4.15)$$

To bound $\Pr\{\mathcal{E}_n\}$, we note that $\mathcal{E}_n$ implies

$$\frac{1}{m}\sum_{j=1}^{n} X_{(i)} = \frac{1}{m}\sum_{j=1}^{n} X_i \le X_{(n)} \le 1.$$

Thus, since $\mathsf{E}[X_i] = a/(a+1)$,

$$\Pr\{\mathcal{E}_n\} \le \Pr\left\{\sum_{j=1}^{n} X_j \le m\right\} \le \Pr\left\{\sum_{j=1}^{n} X_j - n\mathsf{E}[X_i] \le -\left(\frac{an}{a+1} - m\right)\right\}.$$

4.1. ANALYSIS OF CLASSICAL GREEDY HEURISTICS

Then by Theorem 2.5 (page 19) we have, for any fixed m, $\Pr\{\mathcal{E}_n\} = O(e^{-cn})$. Substituting into (4.15) and then introducing the function $g_n = (n+1)^{p/a} d_{n,m}$ yields the recurrence

$$g_n \leq \left(\frac{n+1}{n}\right)^{p/a} \left(\frac{n}{n+p/a}\right) g_{n-1} + O(n^{p/a} e^{-cn}).$$

Now it is readily verified that

$$\left(\frac{n+1}{n}\right)^{p/a} \left(\frac{n}{n+p/a}\right) = 1 + O(n^{-2}),$$

so this recurrence becomes

$$g_n \leq \left(1 + O(n^{-2})\right) g_{n-1} + O(n^{p/a} e^{-cn}).$$

Since $\prod_{n=1}^{\infty} \left(1 + O(n^{-2})\right)$ converges, it follows easily that $g_n = O(1)$, and hence

$$\mathsf{E}[A^{\mathrm{LPT}}(L_n, m)^p] \leq \mathsf{E}[D^{\mathrm{LPT}}(L_n, m)^p] \leq d_{n,m} = O(n^{-p/a}). \blacksquare$$

For the case $p = 1$ a matching lower bound was proved in [CFL84a]. As this result exploits a simple but useful property of the median of a distribution, it is reproduced below. We make use of the following combinatorial result.

Lemma 4.5 *Let x_i, $1 \leq i \leq n$, be sample values of the $X_{(i)}$ (so $x_n \geq x_{n-1} \geq \cdots \geq x_1$). Let v_i denote the total idle time in the LPT schedule for $x_n, x_{n-1}, \ldots, x_i$. Then $v_i \geq |v_{i+1} - x_i|$.*

Proof. Let l_i denote the length of the schedule for $x_n, x_{n-1}, \ldots, x_i$. If $l_i = l_{i+1}$, then clearly $v_i = v_{i+1} - x_i \geq 0$ and the lemma follows. If $l_i > l_{i+1}$, then

$$v_i = v_{i+1} - x_i + m(l_i - l_{i+1}), \tag{4.16}$$

so that if $v_{i+1} \geq x_i$ we are again done. If $v_{i+1} < x_i$, then $l_i - l_{i+1} \geq x_i - v_{i+1}$ is easily seen to hold, so that (4.16) implies

$$v_i \geq m(x_i - v_{i+1}) + v_{i+1} - x_i = (m-1)(x_i - v_{i+1}) \geq |v_{i+1} - x_i|. \blacksquare$$

Theorem 4.6 ([CFL84a]) *If, for some $a > 0$, $F(x) = x^a$ for $0 \leq x \leq 1$, then*

$$\mathsf{E}[D^{\mathrm{LPT}}(L_n, m)] = \Theta(n^{1/a}).$$

Proof. By Theorem 4.4 we need only prove the lower bound $\Omega(n^{1/a})$. From Lemma 4.5 we may write

$$E[D^{\text{LPT}}(L_n,m)] = \frac{1}{m}E[V_1] \geq \frac{1}{m}E\left[|V_2 - X_{(1)}| \mid X_{(2)},\ldots,X_{(n)}\right],$$

where V_i is the random variable with the sample v_i in the lemma. For later convenience we normalize with respect to $X_{(2)}$ to obtain

$$E[V_1] \geq E\left[X_{(2)} E\left[\left|\frac{V_2}{X_{(2)}} - \frac{X_{(1)}}{X_{(2)}}\right| \mid X_{(2)},\ldots,X_{(n)}\right]\right],$$

and then obtain the lower bound

$$E[V_1] \geq E\left[X_{(2)} \min_{\alpha} E\left[\left|\alpha - \frac{X_{(1)}}{X_{(2)}}\right| \mid X_{(2)}\right]\right].$$

But for any random variable Y, $E[|\alpha - Y|]$ is minimized when α is chosen to be the median, med Y, of the distribution. Thus we have the bound

$$E[V_1] \geq E[X_{(2)} h(X_{(2)})], \tag{4.17}$$

where $h(X_{(2)}) = E[|Z - \text{med } Z| \mid X_{(2)}]$ with $Z = X_{(1)}/X_{(2)}$, so that

$$\Pr\{Z \leq x \mid X_{(2)} = y\} = \begin{cases} F(xy)/F(y) & \text{for } 0 \leq x \leq 1, \\ 1 & \text{for } x > 1. \end{cases}$$

For $F(x) = x^a$, $0 \leq x \leq 1$, we have the simplification $F(xy)/F(y) = F(x) = x^a$, so $\Pr\{Z \leq x\} = x^a$, $0 \leq x \leq 1$. A calculation gives

$$h(X_{(2)}) = \frac{a}{a+1}\left(1 - \frac{1}{2^{1/a}}\right). \tag{4.18}$$

Finally, if $F_{(2)}$ denotes the distribution of $X_{(2)}$, then in terms of the gamma function we have

$$E[X_{(2)}] = \int_0^1 [1 - F_{(2)}(x)]dx$$

$$= \int_0^1 (1 - x^a)^n \, dx + n\int_0^1 x^a (1-x^a)^{n-1} \, dx$$

$$\sim \frac{\left(\frac{1}{a} + \frac{1}{a^2}\right)\Gamma\left(\frac{1}{a}\right)\Gamma(n+1)}{\Gamma\left(n+1+\frac{1}{a}\right)} \sim \frac{\left(\frac{1}{a} + \frac{1}{a^2}\right)\Gamma\left(\frac{1}{a}\right)}{n^{1/a}} \quad \text{as } n \to \infty, \tag{4.19}$$

4.2. DIFFERENCING METHODS

where we have used [AS70, Sections 6.1.47, 6.2.1, 6.2.2]. The theorem then follows from (4.17)–(4.19). ∎

Further asymptotic results for the distribution $F(x) = x^a$, $0 \leq x \leq 1$, can be found in [Boxm84, RF86, FR87]. Boxma [Boxm84] shows that the tail probability $\Pr\{A^{\text{LPT}}(L_n, m) > z\}$ tends to 0 exponentially fast in n for any fixed $0 \leq z \leq (m-1)/m$. Frenk and Rinnooy Kan [FR87] prove that, for any fixed m, $A^{\text{LPT}}(L_n, m)$ converges to 0 (a.s.) at a rate $O\big((\log\log n/n)^{1/a}\big)$; more precisely,

$$\limsup_{n \to \infty} \left(\frac{n}{\log \log n}\right)^{1/a} A^{\text{LPT}}(L_n, m) < \infty \text{ (a.s.)}.$$

These papers present analogous results for the exponential distribution $F(x) = 1 - e^{-\lambda x}$, $x \geq 0$. In addition, the results in [RF86, FR87] are extended to multiprocessor systems in which the processors are allowed to have different speeds.

4.2 Differencing methods

Karp and Karmarkar[KK82] have devised an elegant heuristic, called the *differencing method*, that greatly improves on the asymptotic behavior of the LPT heuristic described in the previous section. In this section we sketch the analysis of one efficient application of this heuristic. These methods can most readily be explained for the case $m = 2$, although they can be extended to arbitrary values of m so as to yield similar asymptotic behavior. For the remainder of this section we restrict ourselves to $m = 2$; this is sufficient to cover the main points of the asymptotic analysis. We conclude the section by showing that one of the simpler algorithms that can be viewed as a differencing method does not have good behavior.

If X and Y are two elements in some list, let us use the phrase "differencing X and Y" to mean replacing X and Y by $|X - Y|$. A key observation of the set-differencing method is that differencing X and Y is equivalent to making the decision that they will go into opposite blocks. Thus, if we iterate this operation until we have reduced the set to a single value, we have effectively produced a partition with that value as its block-sum difference. It is not hard to actually reconstruct the partition by working back through the differencing operations that were performed. For example, let $L_n^{(i)}$ denote the set following the ith differencing operation ($L_n^{(0)} = L_n$). Suppose X and Y in $L_n^{(i-1)}$ were differenced in forming $L_n^{(i)}$. Then to obtain the solution

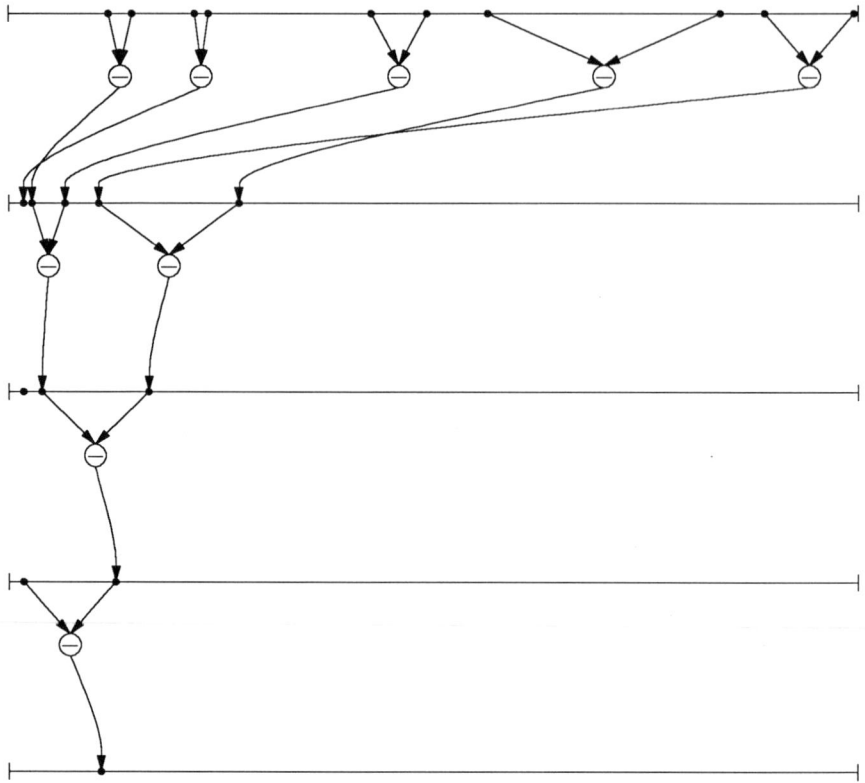

Figure 4.2: Illustration of PDM.

(partition) for $L_n^{(i-1)}$ from the solution for $L_n^{(i)}$, we simply remove the element $|X - Y|$ from the partition for $L_n^{(i)}$ and then append X and Y in the obvious way that preserves the block-sum difference of the partition for $L_n^{(i)}$.

Various algorithms can be obtained by choosing different methods for selecting the items to difference. For example, define the *paired differencing method*, PDM, as follows; see Figure 4.2. Order the items largest to smallest; pair the largest two, the third and fourth largest, etc., possibly leaving the smallest item unpaired; now difference each pair. Call this a *phase* of PDM. Iterate these phases on values remaining until only one value remains. A second possibility, which we call the *largest differencing method*, LDM, is as follows: pick the largest two items and difference them; iterate until a single value remains. Figure 4.3 illustrates this algorithm on the same input data as used in Figure 4.2. At first, PDM is quite appealing, and one might expect

4.2. DIFFERENCING METHODS

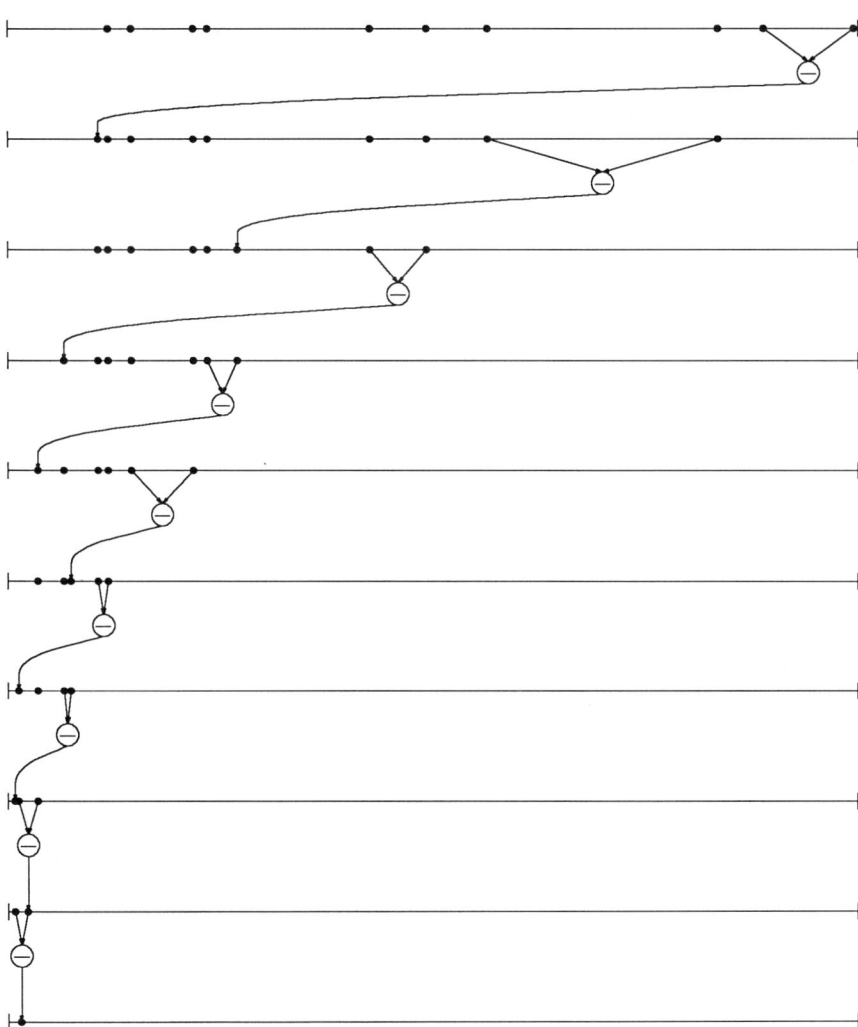

Figure 4.3: Illustration of LDM.

it to yield a final difference of $n^{-\Omega(\log n)}$ by the following heuristic argument. It is clear that there are $\Theta(\log n)$ phases before PDM halts, and in the first half of the phases there will be $\Omega(\sqrt{n})$ items present, so by differencing pairs we might hope to achieve a reduction by a factor of $\sqrt{n}$ or more in the typical sizes of the items remaining. We will see later in this section that this intuition is far from correct. The behavior of LDM is still open to our knowledge. Note that the distribution of the remaining items becomes conditioned in a complicated way as the algorithm proceeds.

These difficulties are handled by Karmarkar and Karp [KK82], who introduce another algorithm DM* that is easier to analyze. DM* to some extent uses the approach of PDM, where paired numbers are usually close in size. However, by a certain randomization of pair selection, a simplified inductive relationship can be proved. A strong result can then be obtained, under the assumption that the density $f(x)$ of the X_i satisfies the following smoothness (Lipschitz) condition: there exists a $\beta > 0$ such that for all x and y in $[0, 1]$, $|f(x) - f(y)| \leq \beta |x - y|$.

Theorem 4.7 ([KK82]) *There exists a $c > 0$ such that with a probability that tends to 1 as $n \to \infty$, the difference in block sums for the DM* partition satisfies* $\text{DM}^*(L_n, 2) \leq n^{-c \log n}$.

After a description of DM* we briefly discuss the proof of this theorem. DM* is organized into phases patterned after those of PDM. The input to the ith phase is a set S_{i-1} of numbers in an interval $[0, \alpha_{i-1}]$, where $S_0 = L_n$ and $\alpha_0 = 1$. The ith phase begins by subdividing the interval $[0, \alpha_{i-1}]$ into subintervals of length α_i, where the number α_{i-1}/α_i of subintervals is some fraction $1/K$ of $|S_{i-1}|$.

Within each subinterval, pairs of values are selected at random until there remains at most one value (the *odd* value, if one exists). The absolute difference of each pair is placed in a set S'_i. (Note that all differences in S'_i are in $[0, \alpha_i]$.) After all subintervals have been processed in this way, the algorithm then applies LDM to the set C of odd values that remain, if C is nonempty, to obtain a value d. The value d is then added to S'_i to form the set S_i. If $d \leq \alpha_i$, then the ith phase terminates. Otherwise, the largest value in S_i (initially d) is iteratively differenced against randomly chosen elements in S_i until all values are in $[0, \alpha_i]$, at which point the ith phase terminates. After the last phase, when a single number remains, the DM* partition of L_n is constructed by a simple backtracking procedure similar to the one mentioned earlier.

4.2. DIFFERENCING METHODS

A primary objective of the analysis is a proof that for each i the numbers within each subinterval of $[0, \alpha_i]$ are statistically independent and approximately uniformly distributed. In Figure 4.4 we note how a distribution that is nearly uniform can be converted to a distribution that is exactly uniform by rejecting a small number of the sample points. If we form a set of points by first drawing according to the density f, but then only accepting a point x with probability $\ell/f(x)$, we effectively reject the portion of the density above the line at height ℓ. Thus the distribution of what remains is uniform. This sort of "resampling," while not a part of the DM* algorithm itself, plays a key role in the analysis; the details of the argument are beyond the scope of this book.

The performance of PDM is much worse than that of DM*. Karp [Karp85] empirically observed that PDM tends to work very poorly, since there is a tendency for a large item to form that cannot be substantially reduced by differencing with the other, much smaller, items. In fact, this can be proved:

Theorem 4.8 ([Luek87]) *With i.i.d. input from $U(0, 1)$, the expected size of the difference produced by PDM is $\Theta(n^{-1})$.*

Proof. The proof exploits properties of the Poisson process. Let D be the final difference achieved when the items to be partitioned are n i.i.d. draws from $U(0, 1)$. Now assume instead that the items are given by the first n epochs $\hat{X}_1, \hat{X}_2, \ldots, \hat{X}_n$ of a Poisson process with rate 1, and let $\hat{D}$ be the difference achieved. As in Section 4.1, by the observations in Section 2.7.1 it is easy to verify that

$$E[D] = E[\hat{D}]/(n+1). \tag{4.20}$$

After the first stage of PDM, there are clearly $\lceil n/2 \rceil$ differences remaining, and they are i.i.d. exponential random variables with mean 1. Hence by Lemma 2.9 (page 34), after the second stage the differences are $z_1, z_2, \ldots, z_m$, where $m = \lceil \lceil n/2 \rceil / 2 \rceil$ and

$$\Pr\{z_i > x\} = e^{-(2i-1)x}. \tag{4.21}$$

Next we establish that if we let L be the largest of the z_i, and let L_1 and L_2 be the largest and second largest elements of $\{z_2, \ldots, z_m\}$, then

$$z_1 - L_1 - L_2 \leq \hat{D} \leq L. \tag{4.22}$$

The upper bound in (4.22) is immediate, since the difference of two items is certainly no more than the maximum of the two. To see the lower bound, consider two cases. On the one hand, if z_1 is not the unique maximum of

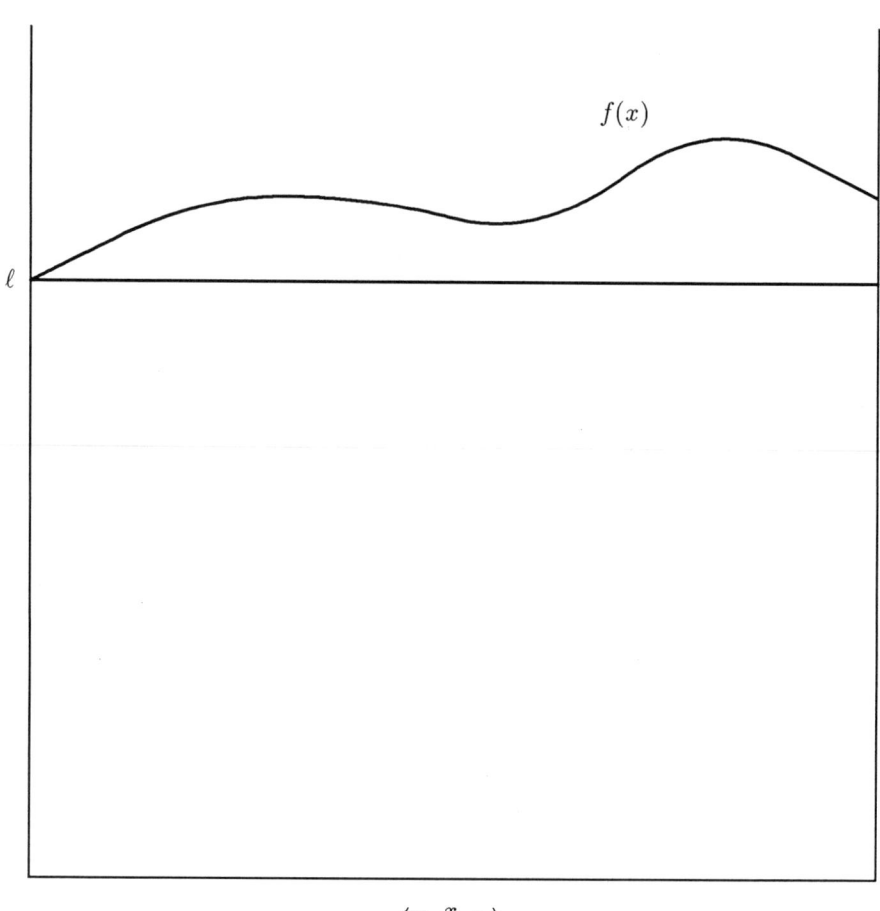

Figure 4.4: Sampling to control distributions.

4.2. DIFFERENCING METHODS

$\{z_1, z_2, \ldots, z_m\}$, then we must have $L_1 = L \geq z_1$, so the lower bound is trivial. On the other hand, if z_1 is the unique maximum of $\{z_1, z_2, \ldots, z_m\}$, then L_1 and L_2 must be, respectively, the second and third largest elements of this set. Hence after the next stage of PDM, the remaining items will be $z_1 - L_1$ and a set of items whose sum is at most L_2, from which the lower bound follows immediately.

Taking expectations in (4.22), we have

$$\mathsf{E}[z_1] - \mathsf{E}[L_1] - \mathsf{E}[L_2] \leq \mathsf{E}[\hat{D}] \leq \mathsf{E}[L]. \tag{4.23}$$

From (4.21) we immediately compute $\mathsf{E}[z_1] = 1$. It remains to estimate the expectations of L, L_1, and L_2. From (4.21) and the definition of L we have

$$\Pr\{L > x\} = \Pr\{\exists i \in \{1, 2, \ldots, m\} : z_i > x\}$$

$$\leq \sum_{i=1}^{m} \Pr\{z_i > x\} = \sum_{i=1}^{m} e^{-(2i-1)x} \leq \sum_{i=1}^{\infty} e^{-(2i-1)x}$$

$$= \frac{e^{-x}}{1 - e^{-2x}}.$$

Hence

$$\Pr\{L > x\} \leq \max\left\{1, \frac{e^{-x}}{1 - e^{-2x}}\right\}.$$

Numerical evaluation and integration yield $\mathsf{E}[L] \leq 1.21$. Similarly, one computes $\mathsf{E}[L_1] \leq 0.52$. Finally, to estimate L_2, we note that

$$\Pr\{L_2 > x\} = \Pr\{\exists i, j \in \{2, 3, \ldots, m\} : i \neq j, \, z_i > x \text{ and } z_j > x\}$$

$$\leq \sum_{2 \leq i < j \leq m} \Pr\{z_i > x\} \Pr\{z_j > x\}$$

$$\leq \sum_{i=2}^{\infty} \sum_{j=i+1}^{\infty} e^{-(2i-1)x} e^{-(2j-1)x}$$

$$\leq \sum_{i=2}^{\infty} \sum_{j=i+1}^{\infty} e^{-(2(i+j)-2)x}.$$

We now count the number of occurrences of each value of the exponent to obtain

$$\Pr\{L_2 > x\} \leq \sum_{i=0}^{\infty} (i+1)\left(e^{-(8+4i)x} + e^{-(10+4i)x}\right)$$

$$= (e^{-8x} + e^{-10x}) \sum_{i=0}^{\infty} (i+1) e^{-4ix}$$

$$= \frac{e^{-8x} + e^{-10x}}{(1 - e^{-4x})^2}.$$

Hence
$$\Pr\{L_2 > x\} \le \max\left\{1, \frac{e^{-8x} + e^{-10x}}{(1 - e^{-4x})^2}\right\}.$$

Numerical evaluation and integration yield $\mathsf{E}[L_2] \le 0.30$. Inserting these estimates into (4.23), we obtain

$$1 - 0.52 - 0.30 \le \mathsf{E}[\hat{D}] \le 1.21,$$

so $\mathsf{E}[\hat{D}] = \Theta(1)$. Then (4.20) yields the theorem. ∎

([Luek87] actually shows the stronger result that there exist positive constants α, β, and ρ such that for large n we have $\Pr\{D \in [\alpha/n, \beta/n]\} \ge \rho$.)

Although the performance of PDM itself is poor, Tsai [Tsai90] points out that a single stage of the Paired Differencing Method can be a useful first step if one wants to balance not only the total workload but also the number of jobs placed on each processor. This single stage of paired differencing does not cause $\Theta(n^{-1})$ behavior; in fact, Tsai shows that if we follow it by an application of LPT, the workload difference is almost surely $O(\log n/n^2)$ if the input are i.i.d. uniform.

4.3 On the optimum solution

The preceding results leave open a number of interesting questions, e.g.,

- Can a polynomial-time algorithm achieve a difference of $O(e^{-\alpha n})$, for some positive α, in expectation or at least with high probability?

- What is the behavior of the true optimum?

While we do not know an answer to the first question, an interesting analysis for the second appears in [KKLO86]. Here we will consider partitioning variables into only two blocks, so a partition is determined by its first block.

Let $\mathcal{A}$ denote the family of all subsets of $N = \{1, 2, \ldots, n\}$. Then the set of partitions is $\{(A, \overline{A})\}_{A \in \mathcal{A}}$; note that in this way of counting partitions we consider $(A, \overline{A})$ and $(\overline{A}, A)$ to be distinct. Say a partition is *even* if $|A| = |\overline{A}|$, i.e., if $|A| = n/2$; we let $\mathcal{A}_e$ be $\{A \in \mathcal{A} : (A, \overline{A}) \text{ is even}\}$. In [KKLO86] the minimum difference achievable is considered both when the partition is restricted to be even and when it is unrestricted. We first consider the case of partitions restricted to be even. One can estimate the distribution of the best partition difference fairly accurately for small arguments. (Note that because of the $O(n^{-1})$ term in (4.24), this theorem does not yield a bound on the

4.3. ON THE OPTIMUM SOLUTION

expectation of the best difference. We do not know whether the expectation is $O(e^{-\alpha n})$ for some $\alpha > 0$.

Theorem 4.9 ([KKLO86]) *Assume that $X_1, X_2, \ldots, X_n$ are i.i.d., and that X_1 has a bounded density, variance $\sigma^2 > 0$, and finite third and fourth moments. Let $G(\epsilon)$ be the probability that the best even partition difference is ϵ or less. Then*

$$G(\epsilon) \leq z\big(1 + O(n^{-1})\big), \text{ and}$$

$$G(\epsilon) \geq \frac{z}{1+z}\big(1 + O(n^{-1})\big), \quad (4.24)$$

where

$$z = \frac{2^n \epsilon}{\sigma \pi n}.$$

Hence the median value of the distribution of the minimum difference of even partitions is $\Theta(n/2^n)$.

Proof. To minimize notation, assume $\sigma^2 = 1$; once the theorem is proved in this case, the generalization to arbitrary $\sigma^2 > 0$ follows immediately by scaling. Also, we first assume assume ϵ is in the range

$$0 < \epsilon \leq n^2 2^{-n}; \quad (4.25)$$

then for larger ϵ the bounds in the theorem follow readily. Let

$$\Delta(A) = \sum_{i \in A} X_i - \sum_{i \in \overline{A}} X_i$$

and

$$Y_n = \Big|\{A \in \mathcal{A}_e : |\Delta(A)| \leq \epsilon\}\Big|, \quad (4.26)$$

i.e., Y_n is a random variable denoting the number of even partitions that achieve a difference of ϵ or less. (Note that Y_n will be an even integer.) Then $G(\epsilon) = \Pr\{Y_n > 0\}$, and we must estimate upper and lower bounds on this probability. Let X_{sym} be the *symmetrization* of X_1, i.e., a random variable distributed as the result of subtracting two independent samples of X_1; note that the variance of X_{sym} is 2. Let g be the density of X_{sym}, and g^{*k} be the density of the kth partial sum of X_{sym}.

An upper bound is rather straightforward. For any nonnegative integer-valued random variable we have the simple bound

$$\Pr\{Y_n > 0\} \leq \mathsf{E}[Y_n].$$

In fact, in the present case the values assumed by Y_n are always multiples of 2, so we can write

$$\Pr\{Y_n > 0\} \leq \mathsf{E}[Y_n]/2. \tag{4.27}$$

The expectation of Y_n is easily calculated; we have

$$\begin{aligned}
\mathsf{E}[Y_n] &= \mathsf{E}\left[|\{A \in \mathcal{A}_e : |\Delta(A)| \leq \epsilon\}|\right] \\
&= \sum_{A \in \mathcal{A}_e} \Pr\{|\Delta(A)| \leq \epsilon\} \\
&= |\mathcal{A}_e| \int_{-\epsilon}^{\epsilon} g^{*n/2}(x)\, dx = \binom{n}{n/2} \int_{-\epsilon}^{\epsilon} g^{*n/2}(x)\, dx,
\end{aligned} \tag{4.28}$$

To estimate the integral, we will apply Theorem 2.2 (page 13) to estimate the value of $g^{*n/2}$. Letting φ denote the characteristic function of X_1, by [Fell71, Section XV.1, Corollary to Lemma 2], the characteristic function for X_{sym} is $|\varphi^2|$. Then, since its density g is bounded and its characteristic function is nonnegative-real-valued, we conclude by [Fell71, Section XV.3, Corollary to Theorem 3] that this characteristic function is integrable. Thus an application of Theorem 2.2, with $\sigma^2 = 2$ and $\mu_3 = 0$, gives the estimate, for $m \in \{1, 2, \ldots, n\}$,

$$g^{*m}(x) = \frac{1}{\sqrt{2m}}\,\mathfrak{n}\!\left(\frac{x}{\sqrt{2m}}\right) + O(m^{-3/2}),$$

and, therefore, in view of (4.25),

$$g^{*m}(x) = \frac{1}{\sqrt{4\pi m}}\bigl(1 + O(m^{-1})\bigr) \quad \text{for } |x| \leq \epsilon. \tag{4.29}$$

Substituting (4.29) into (4.28), one obtains

$$\begin{aligned}
\mathsf{E}[Y_n] &= \binom{n}{n/2} \int_{-\epsilon}^{\epsilon} \frac{1}{\sqrt{2\pi n}}\bigl(1 + O(n^{-1})\bigr)\, dx, \\
&= \binom{n}{n/2} \frac{2\epsilon}{\sqrt{2\pi n}}\bigl(1 + O(n^{-1})\bigr) \\
&= \frac{2^{n+1}\epsilon}{\pi n}\bigl(1 + O(n^{-1})\bigr) = 2z\bigl(1 + O(n^{-1})\bigr),
\end{aligned} \tag{4.30}$$

where we have used

$$\binom{n}{n/2} = \bigl(1 + O(n^{-1})\bigr) 2^{n+1}/\sqrt{2\pi n}, \tag{4.31}$$

4.3. ON THE OPTIMUM SOLUTION

which is a well-known consequence of Stirling's approximation (see [Knut73, Section 1.2.11], or [AS70, formulas 6.1.37–38]). Then by (4.27) we have the upper bound in (4.24).

By means of a suitable lower bound for $\Pr\{Y_n > 0\}$, we would also like to be able to show that good partitions are likely to occur. Unfortunately, the expectation of Y_n does not in itself provide much help. (It is easy to construct examples of a random variable that is extremely unlikely to be greater than 0, even though its expectation is much greater than 1. For example, let Y be 10^6 with probability 10^{-4}, and 0 otherwise.) However, using the second moment method described in Section 2.7.3, we can provide a bound if we also know $\mathsf{E}[Y_n^2]$. As explained there we have

$$\Pr\{Y_n > 0\} \geq \frac{\mathsf{E}[Y_n]^2}{\mathsf{E}[Y_n^2]}, \qquad (4.32)$$

so we can show that Y_n is likely to be nonzero by showing that $\mathsf{E}[Y_n]^2$ is comparable to $\mathsf{E}[Y_n^2]$.

At first, it might seem that the calculation of the second moment of Y_n would be much more difficult than calculating the first moment, but a standard method is available. For any $A \in \mathcal{A}_e$, let

$$Z(A) = 1_{\{|\Delta(A)| \leq \epsilon\}},$$

so $Z(A)$ is one when A yields a partition difference bounded by ϵ, and 0 otherwise. Then, since $Y_n = \sum_{A \in \mathcal{A}_e} Z(A)$, we have

$$\mathsf{E}[Y_n^2] = \mathsf{E}\left[\sum_{A \in \mathcal{A}_e} Z(A) \sum_{B \in \mathcal{A}_e} Z(B)\right] = \sum_{A \in \mathcal{A}_e} \sum_{B \in \mathcal{A}_e} \mathsf{E}[Z(A)Z(B)].$$

Now it is not hard to see that $\mathsf{E}[Z(A)Z(B)]$ depends on A and B only through $|A \cap B|$, so let

$$E_m = \mathsf{E}[Z(A)Z(B)] \text{ when } |A \cap B| = m.$$

Then, by consideration of the number of choices of A and B with various intersection cardinalities,

$$\mathsf{E}[Y_n^2] = \binom{n}{n/2} \sum_{m=0}^{n} \binom{n/2}{m}\binom{n/2}{n/2-m} E_m = \binom{n}{n/2} \sum_{m=0}^{n} \binom{n/2}{m}^2 E_m. \qquad (4.33)$$

E_m is estimated in several cases, which we briefly outline.

a) First suppose that $m = 0$ or $m = n/2$, so $B = A$ or $B = \overline{A}$. Then $|\Delta(A)| = |\Delta(B)|$, so $Z(A) = Z(B)$ and we have

$$E_0 = E_{n/2} = \mathsf{E}[Z(A)Z(B)] = \mathsf{E}[Z(A)]$$
$$= \int_{-\epsilon}^{\epsilon} g^{*n/2}(x)\,dx = \frac{2\epsilon}{\sqrt{2\pi n}}\bigl(1 + O(n^{-1})\bigr),$$

where we have again used the estimate (4.29).

b) If m is near $n/4$, say, $\lceil n/4 - n^{3/5}\rceil \le m \le \lfloor n/4 + n^{3/5}\rfloor$, let $S_{\text{com}} = \frac{1}{2}\bigl(\Delta(A) + \Delta(B)\bigr)$ and $S_{\text{dif}} = \frac{1}{2}\bigl(\Delta(A) - \Delta(B)\bigr)$. (We assume n is large enough so that $n^{3/5} \le n/8$.) It is not hard to verify that S_{com} and S_{dif} are independent with densities g^{*m} and $g^{*(n/2-m)}$, respectively. Moreover, $Z(A)Z(B)$ is 1 if and only if $(S_{\text{dif}}, S_{\text{com}})$ lies in the diamond-shaped region $-\epsilon \le S_{\text{dif}} \pm S_{\text{com}} \le \epsilon$, which has area $2\epsilon^2$. In this region another application of the estimate in (4.29) yields

$$E_m = \iint_{-\epsilon \le x \pm y \le \epsilon} g^{*m}(x) g^{*(n/2-m)}(y)\,dx\,dy$$
$$= \frac{2\epsilon^2}{4\pi\sqrt{m(n/2-m)}}\bigl(1 + O(n^{-1})\bigr),$$

and hence

$$\sum_{m=\lceil n/4-n^{3/5}\rceil}^{\lfloor n/4+n^{3/5}\rfloor} \binom{n/2}{m}^2 E_m$$

$$= \bigl(1 + O(n^{-1})\bigr) \sum_{m=\lceil n/4-n^{3/5}\rceil}^{\lfloor n/4+n^{3/5}\rfloor} \binom{n/2}{m}^2 \frac{\epsilon^2}{2\pi\sqrt{m(n/2-m)}}$$

$$= \bigl(1 + O(n^{-1})\bigr)\frac{\epsilon^2}{2\pi} \sum_{m=\lceil n/4-n^{3/5}\rceil}^{\lfloor n/4+n^{3/5}\rfloor} \binom{n/2}{m}^2 \frac{1}{\sqrt{m(n/2-m)}}$$

$$= \bigl(1 + O(n^{-1})\bigr)\binom{n}{n/2}\frac{2\epsilon^2}{\pi n},$$

(4.34)

where we have deferred the justification of the last step in (4.34) to Lemma 4.10 below.

c) For the remaining cases the contribution to the sum is negligible, from the crude bound $E_m \le 1$ and the fact that $\binom{n/2}{m}/\binom{n/2}{\lfloor n/4\rfloor}$, for $m \notin [\lceil n/4 - n^{3/5}\rceil, \lfloor n/4 + n^{3/5}\rfloor]$, is small enough to swallow polynomials (see (2.5)).

4.3. ON THE OPTIMUM SOLUTION

Inserting the results of these three cases into (4.33), one obtains

$$E[Y_n^2] = \binom{n}{n/2}\left(\frac{4\epsilon}{\sqrt{2\pi n}}\right) + \binom{n}{n/2}\frac{2\epsilon^2}{\pi n}\left(1 + O(n^{-1})\right)$$

$$= (4z + 4z^2)\left(1 + O(n^{-1})\right),$$

where in the last step we again used (4.31). Substituting this and (4.30) into (4.32), one computes

$$\Pr\{Y_n > 0\} \geq \frac{z}{1+z}\left(1 + O(n^{-1})\right),$$

giving the lower bound in (4.24). ∎

Lemma 4.10

$$\sum_{m=\lceil n/4-n^{3/5}\rceil}^{\lfloor n/4+n^{3/5}\rfloor} \binom{n/2}{m}^2 \frac{1}{\sqrt{m(n/2-m)}} = \left(1 + O(n^{-1})\right)\frac{4}{n}\binom{n}{n/2}.$$

Proof. The idea of the proof is to note (as in another example in [Pipp77]) that $\binom{n/2}{m}^2$ becomes more and more tightly concentrated near the middle of the summation as n becomes large, and that in this vicinity $\left(m(n/2-m)\right)^{-1/2}$ is approximately $4/n$. To make this rigorous, note that by Taylor's theorem we can readily compute the (crude) bound

$$1 \leq \frac{1}{\sqrt{1-x^2}} \leq 1 + 2x^2 \quad \text{for} \quad -\frac{1}{2} \leq x \leq \frac{1}{2},$$

and that with some algebraic manipulation this implies[1]

$$\frac{4}{n} \leq \frac{1}{\sqrt{m(n/2-m)}} \leq \frac{4}{n} + \frac{128}{n^3}\left(m - \frac{n}{4}\right)^2 \quad \text{for } \lceil n/4-n^{3/5}\rceil \leq m \leq \lfloor n/4+n^{3/5}\rfloor.$$

Now, since $\sum_{m=0}^{n}\binom{n/2}{m}^2 = \binom{n}{n/2}$, we have

$$\sum_{m=\lceil n/4-n^{3/5}\rceil}^{\lfloor n/4+n^{3/5}\rfloor} \binom{n/2}{m}^2 \frac{1}{\sqrt{m(n/2-m)}} = \frac{128}{n^3}d_n + \frac{4}{n}\sum_{m=\lceil n/4-n^{3/5}\rceil}^{\lfloor n/4+n^{3/5}\rfloor}\binom{n/2}{m}^2$$

$$= \frac{128}{n^3}d_n + \left(1 + O(n^{-1})\right)\frac{4}{n}\binom{n}{n/2}, \quad (4.35)$$

---

[1]Specifically, let $x = 4m/n - 1$ and multiply both sides by $4/n$. Recall that we assume $n^{3/5} \leq n/8$.

where

$$|d_n| \le \sum_{m=\lceil n/4-n^{3/5}\rceil}^{\lfloor n/4+n^{3/5}\rfloor} \binom{n/2}{m}^2 \left(m-\frac{n}{4}\right)^2.$$

Now from the estimate (2.6), we can write this as

$$|d_n| \le \left(1+o(1)\right) \sum_{m=\lceil n/4-n^{3/5}\rceil}^{\lfloor n/4+n^{3/5}\rfloor} \left(2^{n/2}\sqrt{\frac{2}{n\pi/2}}e^{-2(m-n/4)^2/(n/2)}\right)^2 \left(m-\frac{n}{4}\right)^2$$

$$= \left(1+o(1)\right) \sum_{m=\lceil n/4-n^{3/5}\rceil}^{\lfloor n/4+n^{3/5}\rfloor} \left(2^n\frac{4}{n\pi}\right) \left(m-\frac{n}{4}\right)^2 \exp\left(-\frac{8}{n}\left(m-\frac{n}{4}\right)^2\right).$$

At this point we approximate the integral as in (2.8); it is easy to verify that the error term is of smaller order than the sum, so we have

$$|d_n| \le \left(1+o(1)\right) \left(2^n\frac{4}{n\pi}\right) \int_{\lceil n/4-n^{3/5}\rceil}^{\lfloor n/4+n^{3/5}\rfloor} \left(x-\frac{n}{4}\right)^2 \exp\left(-\frac{8}{n}\left(x-\frac{n}{4}\right)^2\right) dx$$

$$= \left(1+o(1)\right) 2^n \frac{4}{n\pi} \int_{-n^{3/5}}^{n^{3/5}} u^2 e^{-(8/n)u^2} du$$

$$= \left(1+o(1)\right) 2^n \frac{4}{n\pi} 2^{-6} \sqrt{2\pi} n^{3/2}$$

$$= \left(1+o(1)\right) 2^{n-4} \sqrt{2n/\pi}. \tag{4.36}$$

Actually, this estimate is more precise than we need. It suffices to note that from (4.36) and (4.31) we have

$$\frac{128}{n^3}|d_n| = O(n^{-5/2}2^n) = O(n^{-1})\frac{4}{n}\binom{n}{n/2},$$

so from (4.35) we have the lemma. ∎

(We note that an alternative approach to proving the lemma could make use of generating functions, which can give exact formulas for expressions such as $\sum_{m=0}^{n/2} \binom{n/2}{m}^2 \left(\frac{4m}{n}-1\right)^2$.)

Next we briefly describe the argument used in [KKLO86] for the case of unrestricted partitions; that is, we redefine

$$Y_n = \left|\left\{A \in \mathcal{A} : \left|\sum_{i\in A} X_i - \sum_{i\in \overline{A}} X_i\right| \le \epsilon\right\}\right|, \tag{4.37}$$

removing the subscript ϵ from $\mathcal{A}$. Note that this means that we can no longer assume $A = \overline{A}$ for $A \in \mathcal{A}$, so we can no longer conclude the last line in (4.28).

4.3. ON THE OPTIMUM SOLUTION

However, if we let $\delta_1, \delta_2, \ldots, \delta_n$ be random variables that assume the values ± 1 with equal probability, and are independent of the X_i and each other, then (4.37) easily yields

$$\mathsf{E}[Y_n] = \sum_{A \in \mathcal{A}} \Pr\left\{\left|\sum_{i \in A} X_i - \sum_{i \in \overline{A}} X_i\right| \le \epsilon\right\} = 2^n \Pr\left\{\left|\sum_{i=1}^{n} \delta_i X_i\right| \le \epsilon\right\}. \quad (4.38)$$

One could again use Theorem 2.2 to estimate this, but we will briefly indicate the alternative approach that is used in [KKLO86], based on characteristic functions. (Note that the proof of Theorem 2.2 that appears in [Fell71, Section XVI.2] itself makes use of characteristic functions.) To illustrate some of the methods, here we briefly sketch the method of calculation of $\mathsf{E}[Y_n]$.

Let h_n be the density function for $\sum_{i=1}^{n} \delta_i X_i$; then the probability within the right side of (4.38) can be rewritten as

$$\int_{-\epsilon}^{\epsilon} h_n(x)\, dx. \quad (4.39)$$

The characteristic function corresponding to the density h_n can easily be expressed. Note that if ϕ is the characteristic function for one of the X_i, then the characteristic function for $\delta_i X_i$ is

$$\varphi(t) = \tfrac{1}{2}\bigl(\phi(t) + \phi(-t)\bigr) = \operatorname{Re} \phi(t),$$

since the real part of the characteristic function is an even function, and the imaginary part is odd. Hence the characteristic function φ_n for the sum of the $\delta_i X_i$ is just

$$\varphi_n(t) = \bigl(\operatorname{Re} \phi(t)\bigr)^n.$$

But now we need to use the characteristic function to estimate the value of (4.39). If we let $\omega_0(x)$ be 1 for $x \in [-\epsilon, \epsilon]$ and 0 elsewhere, then the integral (4.39) is

$$\int_{-\infty}^{\infty} h_n(x) \omega_0(x)\, dx.$$

It turns out that because of the transform methods employed it is preferable to use a smoother function ω that serves approximately the same purpose as ω_0; we omit the details of the choice of this function. If we now let w be the inverse Fourier transform of ω, a form of Parseval's relation states that

$$\int_{-\infty}^{\infty} h_n(x) \omega(x)\, dx = \int_{-\infty}^{\infty} \varphi_n(t) w(t)\, dt.$$

Combining these observations, one can estimate

$$\mathsf{E}[Y_n] = O\left(\frac{2^n \epsilon}{\sqrt{n}}\right),$$

which gives an upper bound on $\Pr\{Y_n > 0\}$. Similar ideas are useful in the computation of $\mathsf{E}[Y_n^2]$, though the details of the calculation become quite involved; again we omit them. The result of an application of (4.32) then establishes that the median difference for the best partition is $\Theta(\sqrt{n}/2^n)$.

Chapter 5

Bin Packing: The Optimum Solution

5.1 Basic algorithms and bounds

One of the first problems posed by the average-case analysis of bin packing was the determination of $E[\mathrm{OPT}(L_n)]$ when the elements of L_n are distributed uniformly over $[0,1]$. An early upper bound for this distribution was provided by Frederickson's analysis [Fred80] of a heuristic; the technique has played a key role in a number of later papers. The idea behind the heuristic is as follows. It is easy to see that the expected value of the sum of the sizes of the ith smallest item and the ith largest item is 1. This suggests the strategy of packing the items in pairs, i.e., packing the largest item with the smallest, the second largest with the second smallest, etc. We call this the *folding* strategy. Of course, this strategy as stated will not be adequate, since it is not enough that the expected sum of the items in a bin be at most 1; the problem is to satisfy the constraints in all cases.

Frederickson's solution was to choose a value of α near 1, pack items greater than α in separate bins, and then apply the folding strategy to the remaining items. However, when the two items paired by the folding strategy did not fit into a single bin, they were packed into separate bins; by choosing α sufficiently smaller than 1, this occurrence could be made unlikely. The proper choice of α had to be

- large enough so that not too many items used single bins because they were greater than α, but
- small enough so that most paired items would fit into a single bin.

Frederickson showed in fact that the choice $\alpha = 1 - n^{1/3}$ enabled one to pack all of the items with an expected number of bins bounded by $n/2 + O(n^{2/3})$.

One trivially sees that at least $n/2$ bins are needed on the average, since this is the expected total size of the items, so Frederickson's results established that $\mathrm{OPT}(L_n) \sim n/2$.

At first it may appear that these results leave little room for improvement, but as we pointed out in Chapter 1, it is often interesting to examine the expected difference between the behavior of a heuristic and the optimum or a lower bound. In particular, for the bin-packing problem, it is of interest to determine the difference between the number of bins used and the total size of the items, or, equivalently, the total amount of unused space in the bins. Frederickson's results showed this value to be $O(n^{2/3})$; this can be tightened to $O(n^{1/2})$ by an algorithm that does not use the α discussed above. Consider the following matching algorithm: MATCH iterates the following procedure until all items are packed. Let S denote the set of items that remain to be packed. MATCH considers a largest item in S, say, x. If $|S| = 1$ or if no remaining item fits with x, i.e., $y + x > 1$ for all $y \in S - \{x\}$, then MATCH puts x into a bin alone. Otherwise, MATCH puts items x and x' into a bin alone, where x' is a largest remaining item other than x such that $x + x' \leq 1$.

A variety of techniques have been used to show that this approach wastes only $O(\sqrt{n})$ space. Knödel [Knöd81] used the Kolmogorov-Smirnov statistic (see Section 2.7.2. Lueker [Luek82], whose work was motivated by the simulation results and conjectures of Bentley, used the relation discussed in Section 2.7.1 between the exponential distribution and the successive order statistics of uniform random draws. This relation facilitated the analysis by making the differences between the successive order statistics of the items independent. Probably the most elegant analysis is that in [Karp82, KLMS84], which reduces the problem to one involving excesses of heads over tails in a sequence of coin flips, as follows.

Since each bin in the packing produced by the algorithm contains either one or two items, it is clear that

$$\mathrm{MATCH}(L_n) = \frac{n+b}{2}, \qquad (5.1)$$

where b is the number of singleton bins, i.e., bins with exactly one item. Let $b' \leq b$ be the number of singleton bins with items larger than $\frac{1}{2}$. There can be at most one singleton bin with an item no larger than $\frac{1}{2}$, so $b \leq b' + 1$. To estimate $\mathsf{E}[b']$, we plot points on the interval $[0,1/2]$ as follows. For each item $x \leq \frac{1}{2}$, plot a $\ominus$ at x. For each item $x > \frac{1}{2}$, plot a $\oplus$ at $1 - x$. By the *excess* at a point z on the interval we mean the number of $\oplus$'s in $[0, z]$ minus the number of $\ominus$'s in $[0, z]$. It is easy to see that an item corresponding to a $\oplus$ can be packed only with an item corresponding to a $\ominus$ to its left (or at the

5.1. BASIC ALGORITHMS AND BOUNDS

same location), so b' is simply the maximum excess at any point on $[0, 1/2]$. Since all sequences of $\oplus$'s and $\ominus$'s are equally likely, b' is equal in distribution to the maximum excess of the number of heads over tails at any point in a sequence of n flips of a fair coin. This quantity is $\Theta(\sqrt{n})$ in expectation, as shown in Section 2.1.3; see (2.30). Then $\mathsf{E}[b] = \Theta(\sqrt{n})$, so by (5.1) we have

Theorem 5.1 *If n items drawn independently from a uniform distribution over $[0, 1]$ are packed by algorithm MATCH, the expected wasted space is*

$$\mathsf{E}[\mathrm{MATCH}(L_n)] - \frac{n}{2} = \Theta(\sqrt{n}).$$

In [CFGR], for n even, an exact formula for $\mathsf{E}[b']$ is obtained and gives the more precise upper bound $\mathsf{E}[\mathrm{MATCH}(L_n)] - \frac{n}{2} \leq \frac{1}{2}\mathsf{E}[b'] + \frac{1}{2}$, where

$$\mathsf{E}[b'] = (-1)^{n/2}\binom{-3/2}{n/2} - \frac{1}{2}(-1)^{n/2}\binom{-1/2}{n/2} - \frac{1}{2}.$$

A simple argument establishes that for i.i.d. uniform item sizes the algorithm MATCH is optimum up to constant factors on the amount of expected wasted space [Luek82, KLMS84]. Let N be a random variable telling the number of items whose size exceeds $1/2$; clearly, no two of these can lie in the same bin, so $\mathrm{OPT}(L_n) \geq N$. Let T be a random variable telling the total size of the items; since each bin has capacity 1, $\mathrm{OPT}(L_n) \geq T$. Now, for any two random variables Y and Z, not necessarily independent,

$$\Pr\{\max(Y, Z) \leq x\} \leq \min\{\Pr\{Y \leq x\}, \Pr\{Z \leq x\}\}.$$

Thus if we define

$$G(x) = \begin{cases} \Pr\{T \leq x\} & \text{if } x < n/2, \\ \Pr\{N \leq x\} & \text{if } x \geq n/2, \end{cases} \quad (5.2)$$

we have $\Pr\{\mathrm{OPT}(L_n) \leq x\} \leq G(x)$. We now estimate the mean of the distribution G.

Theorem 5.2 *For item sizes drawn independently from $U(0, 1)$,*

$$\mathsf{E}[\mathrm{OPT}(L_n)] - \frac{n}{2} \geq (\sqrt{3} - 1)\sqrt{\frac{n}{24\pi}} + o(\sqrt{n}).$$

Proof. We show that the lower bound claimed in the theorem is $E[X] - n/2$, where X is a random variable with the distribution $G(x)$. Note that

$$E[X] = \int_0^n x \, dG(x)$$

$$= \frac{n}{2} + \int_0^n \left(x - \frac{n}{2}\right) dG(x)$$

$$= \frac{n}{2} + \int_0^{n/2} \left(x - \frac{n}{2}\right) d\Pr\{T \le x\} + \int_{n/2}^n \left(x - \frac{n}{2}\right) d\Pr\{N \le x\}. \tag{5.3}$$

We estimate the two integrals on the right of (5.3) separately. For the first, let $\sigma^2 = 1/12$ denote the variance of the uniform distribution over $[0,1]$, and let $f_n(x)$ be the density function of the centered[1] sum of n i.i.d. samples from this distribution. Then

$$\int_0^{n/2} \left(x - \frac{n}{2}\right) d\Pr\{N \le x\} = -\int_0^{n/2} x f_n(-x) \, dx$$

$$= -\left(1 + o(1)\right) \int_0^{n^{2/3}} x f_n(-x) \, dx,$$

where the last step follows from a standard application of a Hoeffding bound (Theorem 2.5 (page 19)). By applying Theorem 2.2 (page 13) and noting that the third central moment of the uniform distribution vanishes because of symmetry, this becomes

$$\left(-1+o(1)\right) \int_0^{n^{2/3}} x \left(\frac{1}{\sigma\sqrt{n}} \mathfrak{n}\left(\frac{x}{\sigma\sqrt{n}}\right) + O(n^{-3/2})\right) dx,$$

$$= \left(-1 + o(1)\right) \left(\int_0^{n^{2/3}} \frac{x}{\sigma\sqrt{n}} \mathfrak{n}\left(\frac{x}{\sigma\sqrt{n}}\right) dx + O(n^{-3/2}) \int_0^{n^{2/3}} x \, dx\right)$$

$$= \left(-1 + o(1)\right) \left(\sqrt{\frac{n}{24\pi}} + o(1)\right). \tag{5.4}$$

For the second integral on the right of (5.3), we have

$$\int_{n/2}^n \left(x - \frac{n}{2}\right) d\Pr\{N \le x\} = \sum_{i=\lceil n/2 \rceil}^n \left(i - \frac{n}{2}\right) \Pr\{N = i\}.$$

---

[1] Recall from Section 2.1.1 that this means the sum is translated to have mean 0.

5.1. BASIC ALGORITHMS AND BOUNDS

By applying (2.5), then (2.6), and finally (2.7), this becomes

$$\left(1+o(1)\right) \sum_{i=\lceil n/2\rceil}^{\lceil n/2+n^{3/5}\rceil} \left(i-\frac{n}{2}\right) \Pr\{N=i\}$$

$$= \left(1+o(1)\right) \sum_{i=\lceil n/2\rceil}^{\lceil n/2+n^{3/5}\rceil} \left(i-\frac{n}{2}\right) \sqrt{\frac{2}{n\pi}} e^{-2(i-n/2)^2/n}$$

$$= \left(1+o(1)\right) \int_{\lceil n/2\rceil}^{\lceil n/2+n^{3/5}\rceil} \left(x-\frac{n}{2}\right) \sqrt{\frac{2}{n\pi}} e^{-2(x-n/2)^2/n}\, dx.$$

Note that extending the range of integration to $[n/2, \infty]$ introduces a relative error covered by the $o(1)$ term. By letting $u = \sqrt{2/n}(x-n/2)$, simplification yields

$$\left(1+o(1)\right) \int_0^\infty \sqrt{\frac{n}{2\pi}} u e^{-u^2}\, du = \left(1+o(1)\right) \sqrt{\frac{n}{8\pi}}. \tag{5.5}$$

Adding the results of (5.4) and (5.5) gives the lower bound of the theorem. ∎

We remark that [CFGR] shows how a stronger argument can lead to a lower bound with a slightly improved constant:

$$\mathrm{E}[\mathrm{OPT}(L_n)] - \frac{n}{2} \geq \sqrt{n/32\pi} + o(\sqrt{n}).$$

Combining Theorems 5.2 and 5.1, we have:

Theorem 5.3 *For item sizes drawn uniformly from $U(0,1)$,*

$$\mathrm{E}[\mathrm{OPT}(L_n)] - \frac{n}{2} = \Theta(\sqrt{n}).$$

The uniform distribution that we have been assuming is a very special case, of course. Interesting analyses of optimum behavior under more general assumptions have also appeared. We can easily generalize the upper bound portion of Theorem 5.3 to hold for a larger class of distributions. We say that a random variable X and its distribution function F are *symmetric about a* if $X - a$ and $a - X$ have the same distribution. In terms of the distribution F of X, this means that for all z

$$F(a+z) = \Pr\{X - a \leq z\} = \Pr\{a - X \leq z\}$$
$$= \Pr\{X \geq a - z\} = 1 - \Pr\{X < a - z\}$$
$$= 1 - F((a-z)^-),$$

or, equivalently,

$$F(z) + F((2a-z)^-) = 1. \tag{5.6}$$

Corollary 5.4 ([Knöd81]) *For item sizes drawn independently from a distribution F symmetric about 1/2,*

$$\mathsf{E}[\mathrm{OPT}(L_n)] - \frac{n}{2} = O(\sqrt{n}).$$

Proof. Suppose we draw n i.i.d. values $U_1, U_2, \ldots, U_n$ uniformly from [0,1], and set $X_i = F^{-1}(U_i)$. Then the expected number of bins needed to pack the U_i is $n/2 + O(\sqrt{n})$ from the theorem. Also, the X_i are i.i.d. with distribution F. Thus we will be finished if we can show that the number of bins required to pack the X_i is at most the number required to pack the U_i.

Note that MATCH never places more than two items together into a bin. Let u and u' be two values that are placed together into a bin, and let $x = F^{-1}(u)$ and $x' = F^{-1}(u')$. We will show that x and x' will also fit together into a single bin, so the packing of the U_i immediately yields a packing of the X_i using the same number of bins. Suppose instead that x and x' do not fit together. Then

$$1 < x + x' = x + F^{-1}(u') = x + \min\{z: F(z) \geq u'\},$$

so $\min\{z: F(z) \geq u'\} > 1 - x$, and thus $F(1-x) < u'$. Therefore,

$$1 = F(x) + F\big((1-x)^{-}\big) \leq F(x) + F(1-x) < u + u' \leq 1,$$

contradicting the symmetry of F about 1/2. ∎

Note that we cannot in general conclude that $\mathsf{E}[\mathrm{OPT}(L_n)] - \frac{n}{2} = \Omega(\sqrt{n})$ for item sizes drawn from a distribution F symmetric about 1/2. To see this, let F be a single atom at 1/2; then, clearly, F is symmetric about 1/2, but $\mathsf{E}[\mathrm{OPT}(L_n)] - \frac{n}{2} \leq 1/2$, since we never leave more than one-half of one bin empty.

5.2 Perfect packings

Suppose the values in L_n are i.i.d. with distribution function F. Karmarkar defines the *optimum packing ratio* for F to be

$$\lim_{n \to \infty} \frac{\mathsf{E}[\mathrm{OPT}(L_n)]}{\mathsf{E}[\sum_{i=1}^{n} X_i]}.$$

Say that F *allows a perfect packing* (or that F is *perfectly packable*) if its optimum packing ratio is 1. Since the distribution uniform over $[a, b]$ is of special interest, following Karmarkar [Karm82] we will say that the interval

5.2. PERFECT PACKINGS

$[a, b]$ allows a perfect packing if the distribution uniform over that interval does. Thus the results of the previous section enable us to conclude that if $0 \le a \le \frac{1}{2}$ and $b = 1 - a$, then $[a, b]$ allows perfect packing. A question attributed to Karp in [Karm82] is: which intervals allow perfect packings?

Note that the optimum packing ratio is simply $c/\mathsf{E}[X]$ where c is the packing constant for F. (Recall that, informally, the *packing constant* is the limiting expected number of bins used per item in an optimum packing; for the formal definition see Section 2.7.2, page 35.) Hereafter, instead of describing perfect packings by refering to optimum packing ratios with the value 1, we shall refer to packing constants with the value $\mathsf{E}[X]$.

When $a = 0$, i.e., the distribution is of the form $U(0, b)$ for $0 < b \le 1$, the interval always allows a perfect packing. In fact, one can show a stronger result:

Theorem 5.5 ([Knöd81, Karm82, Karp82, Loul84b])[2] *Suppose F is a distribution over $[0, 1]$ that possesses a nonincreasing density function f. Then*

$$\mathsf{E}[\mathrm{OPT}(L_n)] - \mathsf{E}\left[\sum_{i=1}^{n} X_i\right] = O(\sqrt{n}), \tag{5.7}$$

and hence F allows a perfect packing.

Proof. We begin by showing that any decreasing function f can be decomposed into a series of distributions symmetric about inverse powers of 2. Let

$$h_0(x) = \begin{cases} f(1-x) & \text{for } x \in [0, 1/2], \\ f(x) & \text{for } x \in [1/2, 1], \end{cases}$$

and

$$f_1(x) = f(x) - h_0(x);$$

note that $h_0(x)$ is symmetric about $1/2$, and $f_1(x)$ is decreasing over $[0, 1/2]$ and 0 on $[1/2, 1]$. Iterating, we may express f as

$$f(x) = \sum_{k=0}^{\infty} h_k(x),$$

where h_k is symmetric about 2^{-k-1}. By scaling, we can write

$$f(x) = \sum_{k=1}^{\infty} p_k f_k(x),$$

---

[2] The key idea, of decomposing f into symmetric densities, was already present in [Knöd81], although it was not sufficiently developed there to give this theorem. The fact that, under the conditions of the theorem, F allows a perfect packing was proved in [Karm82, Loul84b]. The bound (5.7) is from [Karp82].

where we have chosen p_k and f_k so that $h_k = p_k f_k$ and f_k is a density function. Hence the process of generating the X_i may be viewed as taking place in two steps: first, we select a value of k, which we will call the *type* of X_i, with each value chosen with probability p_k; then we select the value of X_i according to the density f_k.

We begin by packing type-k items into subbins of size 2^{-k}. From Corollary 5.4 (suitably scaled), it follows that we can pack N items distributed as f_k into subbins of size 2^{-k} with an expected wasted space bounded by $\alpha 2^{-k}\sqrt{N}$, where α is some constant independent of k. Since this bound is concave, and since the expected number of items of type k is np_k, by Jensen's inequality we have that the expected wasted space used in packing items of type k into bins of size 2^{-k} is bounded by $\alpha 2^{-k}\sqrt{np_k}$. Thus the expected total amount of wasted space in the subbins is bounded by

$$\sum_{k=0}^{\infty} \alpha 2^{-k}\sqrt{np_k} \leq \alpha\sqrt{n}\sum_{k=0}^{\infty} 2^{-k} = 2\alpha\sqrt{n}. \tag{5.8}$$

Finally, we can treat these subbins as items and pack them into unit-size bins. If we use NFD (defined in Section 1.4.2), the amount of wasted space introduced in this stage is less than 1 since the subbin sizes are inverse powers of 2. Adding this to the bound in (5.8), we see that the expected total wasted space is $O(\sqrt{n})$. ∎

Another result of this type appeared in Section 2.5. In the present terminology, that section proved:

Theorem 5.6 *If $[a,b]$ is contained in $[0,1]$ and symmetric about $1/p$, for some integer p, then $[a,b]$ allows a perfect packing.*

Next we consider more general $[a,b]$. Motivation for the approach can be obtained by examining the dual of the linear relaxation of the integer program in (2.43). (Recall that the problem instance was assumed to have m_j items of size s_j for $j = 1, 2, \ldots, N$; $\mathcal{C}$ was the set of all possible configurations, with $M = |\mathcal{C}|$; and for each $j = 0, 1, \ldots, N$ and $k = 1, 2, \ldots, M$, C_{kj} was the number of items of size s_j in the kth configuration.) This dual is: find the u_j that will

$$\text{maximize } \sum_{j=1}^{N} m_j u_j$$

$$\text{subject to } \sum_{j=1}^{N} u_j C_{kj} \leq 1 \quad \text{for } k = 1, \ldots, M,$$

$$u_j \geq 0 \quad \text{for } j = 1, \ldots, N.$$

5.2. PERFECT PACKINGS

By recalling the definition of the C_{kj}, the constraint in this optimization problem is simply that if a set of items fits in a bin, then the corresponding u values sum to at most 1. We wish to maximize the sum of the u values corresponding to the items (including multiplicities).

This suggests the following approach to bounding packing constants. Say that a function $u : [0,1] \to [0,1]$ is *dual feasible* if for all finite sequences $x_1, x_2, \ldots, x_k$ of positive reals,

$$\sum_{i=1}^{k} x_i \leq 1 \implies \sum_{i=1}^{k} u(x_i) \leq 1.$$

(See Figure 5.1 for an example of such a function.) One easily establishes:

Lemma 5.7 *If u is dual feasible, then the packing constant c for items independently distributed as a random variable X satisfies*

$$c \geq \mathsf{E}[u(X)].$$

(Hence the optimum packing ratio is at least $\mathsf{E}[u(X)]/\mathsf{E}[X]$.)

Proof. Since u is dual feasible, we know that the sum of $u(x)$ over the items x packed in any bin is no more than 1. Thus, for any packing, the number of bins used is bounded below by the sum of $u(x)$ over the sizes x of all the items; the lemma follows easily. ∎

We remark in passing that this function u is similar to the weighting functions that have been used in worst-case analyses of bin packing; see [Coff82] for more information about such analyses.

Of course, a problem that arises in applying Lemma 5.7 is to find good choices for the function u. The identity function provides a trivial example of a dual-feasible function, but it yields a trivial lower bound. It turns out that the best choice of u depends on the interval in question. Computer solutions of various examples were used to guide the selection of u in [Luek83]; an example of a function u for a specific interval is shown in Figure 5.1. For that particular example, using Theorem 5.6, we can asymptotically pack an average of four items of I_1 per bin,[3] three items of I_2 per bin, and two items of I_3 per bin. It is not hard to see that over each of these intervals, the average value of $u(x)$ is equal to the asymptotic expected number of bins required per item, so the bound given by this u function can be seen to be tight. This sort of analysis can be used to compute the packing constant in all of the shaded diamonds in Figure 5.2. (Inside the large upper triangle, the packing constant obviously exceeds $\mathsf{E}[X]$.)

---

[3]That is, the expected number of items in I_1 over the expected number of bins used to pack I_1 approaches 4 for large n.

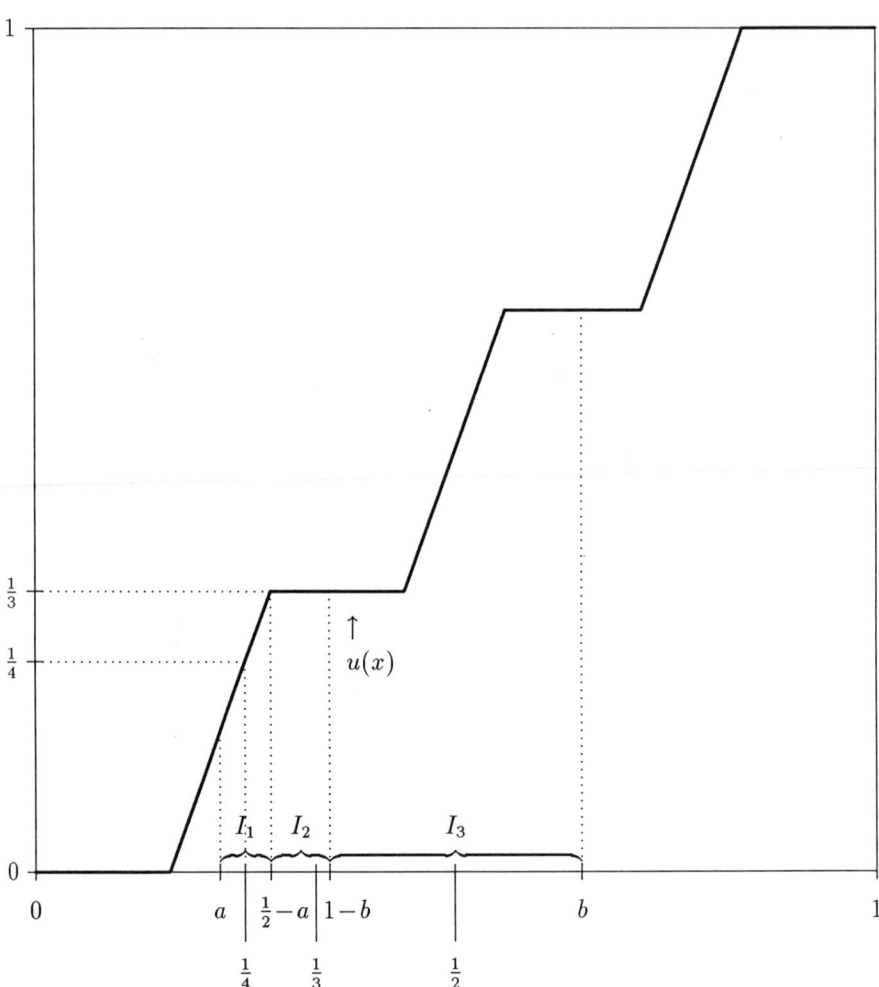

Figure 5.1: An example of a dual-feasible function. Here $[a, b] = [0.22, 0.65]$.

5.2. PERFECT PACKINGS

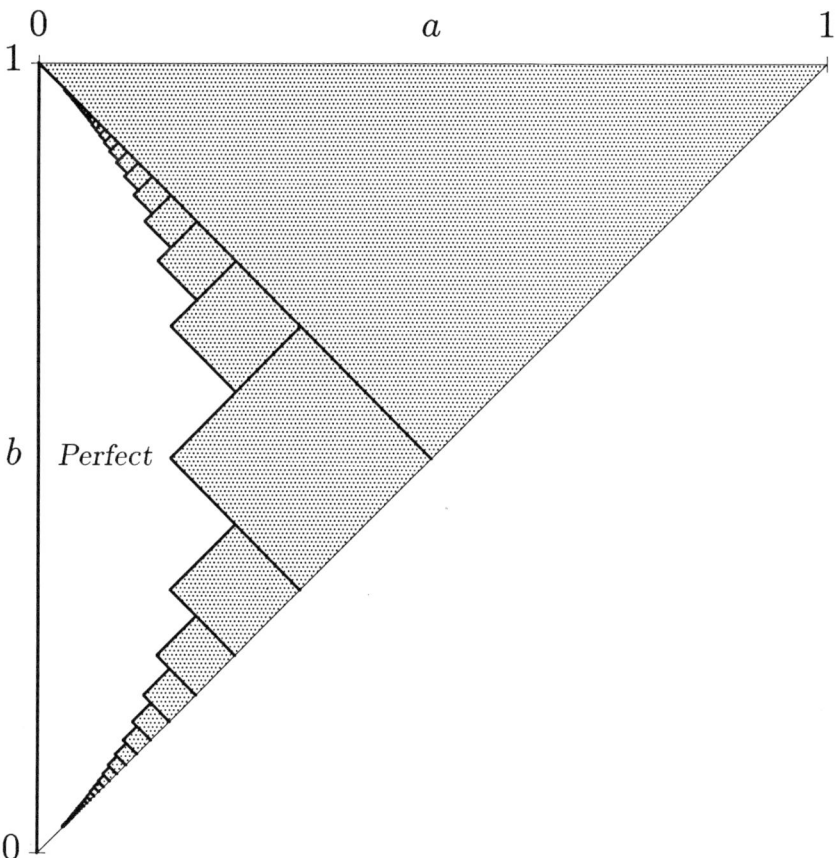

Figure 5.2: Bin packing with items drawn uniformly from $[a, b]$. The packing constant is greater than $\mathsf{E}[X]$ inside the shaded area. Along the bold lines the packing constant equals $\mathsf{E}[X]$. By results to be discussed in Section 5.3, the packing constant is equal to $\mathsf{E}[X]$ inside the unshaded area labeled "Perfect."

5.3 Functional analysis of the packing constant

In this section we explore further the way in which the packing constant c depends on the distribution. For consistency with the relevant literature, we describe random variables by probability measures rather than distribution functions. We consider the nature of the functional $c(\mu)$ that gives the packing constant for a probability measure μ. Rhee and Talagrand [Rhee88, RT88b, RT89e, RT89f] have pioneered the application of notions of topology and functional analysis to this problem.

Lemma 5.7 is motivated by the duality theory of linear programming, and one might, therefore, suspect that a converse to it could be proved. This is more difficult than one might first expect, because the argument based on the duality theorem of linear programming does not seem to apply directly to other than discrete random variables. Rhee and Talagrand [RT89e, RT89f] have, however, been able to prove the following result, by applying the geometric form of the Hahn-Banach theorem. (Our terminology and proof are somewhat different from those of Rhee and Talagrand.)

Theorem 5.8 ([RT89e, RT89f]) *For any probability measure μ, there is a dual-feasible u such that*

$$c(\mu) = \int u(x)\mu\{dx\} = \mathsf{E}[u(X)], \tag{5.9}$$

where X is distributed according to μ.

Before giving the proof, we provide an overview and motivation. Assume a fixed μ. The proof will start by showing that we can come up with a linear mapping l from measures to reals, such that we have

a) for any probability measure ν, $l(\nu) \leq c(\nu)$, and

b) for the particular μ under consideration, $l(\mu) = c(\mu)$.

Given a linear mapping l on a finite dimensional vector space, one can always always find a vector u_l such that the function $l(v)$ is just the inner product $u_l \cdot v$. In fact, by the Riesz representation theorem (see [Sche71, Section II.1]), this same sort of result holds in a much more general context, namely, for bounded functionals on a Hilbert space. This gives hope that we may be able to convert the mapping l into a function u of the sort needed in (5.9).

To formally establish the existence of l, we will use the Hahn-Banach theorem (see [Sche71, Section II.2]). A few definitions will be useful. Recall

5.3. FUNCTIONAL ANALYSIS OF THE PACKING CONSTANT

that a mapping from a vector space to the reals is called a *functional*. A functional l is *linear* if $l(x+y) = l(x) + l(y)$, and $l(\alpha x) = \alpha l(x)$ for arbitrary real α. A functional p is *sublinear* if $p(x+y) \le p(x) + p(y)$, and $p(\alpha x) = \alpha p(x)$ for arbitrary positive α.

Theorem 5.9 (Hahn-Banach) *Let M be a subspace of a given vector space V. Let p be a sublinear functional defined on V, and let f be a linear functional defined on M such that f is bounded above by p. Then f can be extended to a linear functional l on all of V such that l is bounded above by p.*

Proof of Theorem 5.8. The idea of the proof is to let the packing constant play the role of p in the Hahn-Banach theorem, but there is a difficulty: the set of probability measures do not form a vector space. However, the set of all measures of the form

$$a\nu_{\text{pos}} - b\nu_{\text{neg}}, \tag{5.10}$$

where a and b are nonnegative reals and ν_{pos} and ν_{neg} are probability measures, does form a vector space, and clearly includes all the probability measures. So, we choose this set to play the role of V. To apply the theorem, we must generalize the packing constant to a sublinear functional on all of V. The generalization to arbitrary positive measures is very natural: for $\alpha > 0$ let $c(\alpha \nu) = \alpha c(\nu)$. Now any measure $\nu \in V$ can be uniquely decomposed as in (5.10) in such a way that $a+b$ is minimized; we let $c(\nu) = ac(\nu_{\text{pos}}) + bc(\nu_{\text{neg}})$. It is not hard to verify that this functional is sublinear. For M in the Hahn-Banach theorem we simply choose the vector space $\{\alpha\mu\}$ generated by μ, and we define $f(\alpha\mu) = \alpha c(\mu)$, which is clearly linear and bounded by c. Now we can invoke the Hahn-Banach theorem[4] to state that there is a linear functional l such that for all positive measures ν,

$$l(\nu) \le c(\nu), \tag{5.11}$$

and for the particular μ we have fixed,

$$l(\mu) = c(\mu). \tag{5.12}$$

What remains is to see how to produce the desired dual-feasible function u from l. Unfortunately, we see no immediate way to obtain this, so we have

---

[4]Note that we use the Hahn-Banach theorem itself, instead of its geometric form. The technique given here to construct l is somewhat reminiscent of the proof of the geometric form of the Hahn-Banach theorem (see [Sche71, Section VII.3]), with the Minkowski functional replaced by the packing constant.

to do a bit of work. Let $u(x)$ be defined by

$$u(x) = \sup\{l(\nu) : \nu \text{ is a probability measure on } [0, x]\}. \tag{5.13}$$

Note that as an immediate consequence we have

$$\nu \text{ is a probability measure on } [0, x] \implies u(x) \geq l(\nu). \tag{5.14}$$

In fact, with the notation

$$u(x^+) = \lim_{z \downarrow x} u(z) \quad \text{and} \quad u(x^-) = \lim_{z \uparrow x} u(z),$$

we can also conclude that

$$\nu \text{ is a probability measure on } [0, x) \implies u(x^-) \geq l(\nu); \tag{5.15}$$

this follows easily from the observation that for any $\delta > 0$ we can pick $x' < x$ such that $\nu\big((x', x)\big) < \delta$.

Now we show that u is dual-feasible. Suppose that we are given $x_i \in [0, 1]$, $1 \leq i \leq k$, such that $\sum_{i=1}^{k} x_i = 1$. For $1 \leq i \leq k$, let ν_i be an arbitrary probability measure on $[0, x_i]$, and set $\nu = k^{-1} \sum_{i=1}^{k} \nu_i$. It is not hard to see that $c(\nu) \leq 1/k$, since if we draw one item according to each of the ν_i, these items will always fit together in a bin. Hence, using (5.11), we have

$$k^{-1} \geq c(\nu) \geq l(\nu) = l\left(\sum_{i=1}^{k} k^{-1} \nu_i\right) = k^{-1} \sum_{i=1}^{k} l(\nu_i). \tag{5.16}$$

Taking the supremum over all choices of the ν_i and using (5.13), we obtain

$$k^{-1} \geq k^{-1} \sum_{i=1}^{k} u(x_i);$$

multiplying both sides by k, we see that u is dual feasible.

Finally, we must show that u satisfies $\int u(x)\mu\{dx\} = c(\mu)$. In view of (5.12), this means we must show that $\int u(x)\mu\{dx\} = l(\mu)$. Now from (5.13) we see that u must be a (weakly) increasing function from $[0, 1]$ to $[0, 1]$; thus it is continuous except for a countable number of jump discontinuities. Given any $\epsilon > 0$, we can choose $0 = x_0, x_1, \ldots, x_k = 1$ such that if we let I_i be the open interval (x_{i-1}, x_i), then for any $1 \leq i \leq k$, u varies by no more than ϵ over any I_i. Define $\mu_{x_i}(S) = \mu(S \cap \{x_i\})$, and $\mu_{I_i}(S) = \mu(S \cap I_i)$, so that

$$\mu = \sum_{i=0}^{k} \mu_{x_i} + \sum_{i=1}^{k} \mu_{I_i}. \tag{5.17}$$

5.3. FUNCTIONAL ANALYSIS OF THE PACKING CONSTANT

Now, for a positive measure ν on $[0,1]$, let $\|\nu\|$ be defined as $\mu([0,1])$, i.e., the measure of the entire interval $[0,1]$. Then

$$\int u(x)\mu_{x_i}\{dx\} = u(x_i)\|\mu_{x_i}\| \geq l(\mu_{x_i}), \tag{5.18}$$

by (5.14) and the linearity of l. Also, by the choice of the x_i, for $1 \leq i \leq k$ we have

$$\int u(x)\mu_{I_i}\{dx\} = \int_{x_{i-1}}^{x_i} u(x)\mu_{I_i}\{dx\}$$
$$\geq u(x_{i-1}^+)\|\mu_{I_i}\|$$
$$\geq \left(u(x_i^-) - \epsilon\right)\|\mu_{I_i}\|$$
$$\geq l(\mu_{I_i}) - \epsilon\|\mu_{I_i}\|, \tag{5.19}$$

where we have used (5.15). Combining (5.17), (5.18), (5.19), and the linearity of l, and noting that the sum of the total variance of the μ_{x_i} and μ_{I_i} is just 1, we have

$$\int u(x)\mu\{dx\} \geq l(\mu) - \epsilon.$$

Using the fact that ϵ can be chosen arbitrarily small, and using (5.12), we have

$$\int u(x)\mu\{dx\} \geq l(\mu) = c(\mu). \tag{5.20}$$

Lemma 5.7 and (5.20) then combine to yield

$$\int u(x)\mu\{dx\} = c(\mu). \qquad \blacksquare$$

In fact, it is not hard to see that Theorem 5.8 remains true even if we impose a slightly stronger condition on u, namely, that it is a subadditive function from $[0,1]$ to $[0,1]$. Rhee and Talagrand [RT89e] also show that for an arbitrarily small ϵ, we can add the restriction that u be *continuous* and then find a u that satisfies

$$c(\mu) - \epsilon \leq \int u(x)\mu\{dx\} = \mathsf{E}[u(X)];$$

This is related to Theorem 5.11 below.

Let $\mathcal{C}$ be the class of probability measures that allow perfect packing. The following is easily established:

Lemma 5.10 ([Loul84b, Rhee88]) *The set $\mathcal{C}$ is convex. (That is, if μ and μ' are in $\mathcal{C}$, then for all $0 < \alpha < 1$ we also have $\alpha\mu + (1-\alpha)\mu' \in \mathcal{C}$.)*

Rhee [Rhee88] and Rhee and Talagrand [RT89e, RT88b] investigated the class of probability measures for which a perfect packing is possible, and proved some very general results. Here we will state two theorems without proof, and then briefly indicate how they can be used to prove strong results about perfect packings.

In order to describe some of the techniques used by Rhee and Talagrand we briefly review certain topological notions, which we present here in the context of probability measures over [0,1]. A sequence μ_i of probability measures on [0,1] is said to *converge weakly* to μ if for all continuous functions f on [0,1], $\int f(x)\,d\mu_i(x)$ converges to $\int f(x)\,d\mu(x)$; we call μ the *limit* of the μ_i. If a set D of probability measures has the property that for every weakly convergent sequence in D, the limit is also in D, then D is said to be *weakly closed*.

Theorem 5.11 ([Rhee88]) *The set $\mathcal{C}$ is weakly closed.*

Now suppose $k \geq 1$ and let R_k be the set of k-tuples $(x_1, x_2, \ldots, x_k)$ such that $0 \leq x_i \leq 1$ and $\sum_{i=1}^{k} x_i = 1$. Let M_k denote the set of all probability measures on R_k. For $\nu \in M_k$, $\hat{\nu}$ denotes the probability measure determined by first

a) drawing a tuple according to ν and then

b) selecting a single component of this tuple (with each component being equally likely to be selected).

Finally, let $\mathcal{B}_k$ be the family of probability measures obtainable in this way, i.e.,

$$\mathcal{B}_k = \{\hat{\nu}\}_{\nu \in M_k}.$$

For completeness, let $\mathcal{B}_0$ contain only the probability measure concentrated at 0. (Note that $\mathcal{B}_1$ contains only the probability measure concentrated at 1, and that $\mathcal{B}_2$ is the set of probability measures symmetric about 1/2.) Finally, let $\mathcal{B}$ be the class of all probability measures obtainable as (countable) positive linear combinations of probability measures chosen from the $\mathcal{B}_k$. Then the following holds.

Theorem 5.12 ([Rhee88]) *A probability measure μ allows perfect packing if and only if $\mu \in \mathcal{B}$.*

This has the following useful corollary, which we first state formally and then motivate more informally.

5.3. FUNCTIONAL ANALYSIS OF THE PACKING CONSTANT

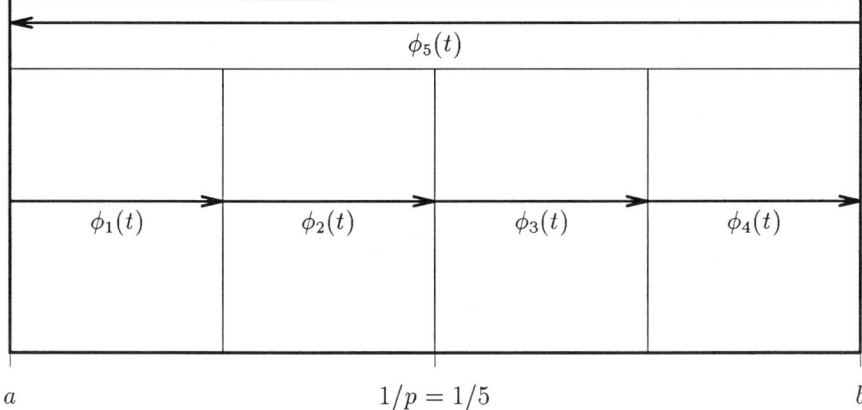

Figure 5.3: Illustration of a simple proof that $U(a,b)$ allows perfect packing if $[a,b] \subseteq [0,1]$ is symmetric about $1/p$. In this illustration $p = 5$.

Corollary 5.13 ([RT88b]) *Suppose we are given an integer $p \geq 2$ and continuous functions $\phi_i : [0,1] \to [0,1]$, $i = 1, 2, \ldots, p$, such that*

$$\bigl(\phi_1(t), \phi_2(t), \ldots, \phi_p(t)\bigr) \in R_p. \tag{5.21}$$

Define a distribution F by

$$F(x) = \frac{1}{p} \sum_{i=1}^{p} \int_0^1 \bigl(1_{\{\phi_i(t) \leq x\}}\bigr) \, dt.$$

Then F allows perfect packing.

Informally, the value of $F(x)$ is the probability that $\phi_i(t) \leq x$ if we pick i uniformly in $\{1, 2, \ldots, p\}$ and t uniformly in $[0,1]$. To state this in picturesque language, we can imagine that we are painting the line segment $[0,1]$ using a spray can that can be positioned at any point and that deposits paint at that point at a fixed rate. For $i = 1, 2, \ldots, p$, we move the paint can according to $\phi_i(t)$, where t progresses at a unit rate from 0 to 1. Then the density $f(x)$ of the distribution $F(x)$ is proportional to the thickness of the paint at point x.

As an example of the power of Corollary 5.13, we will now informally outline a very simple proof, from [RT88b], of Theorem 5.6. See Figure 5.3, where an arrow corresponds to a ϕ_i. We are at the tail of the arrow at time 0, and at the head at time 1. As time progresses from 0 to 1, we move across the arrow at a constant speed. Since the tail positions of the arrows sum to 1, and the head positions sum to 1, it easily follows that at any point in time the sum of our locations on the arrows sums to 1, so (5.21) holds. The height of the box containing an arrow indicates the thickness of the paint deposited. Note that the speed with which we traverse an arrow must be proportional to its length, and hence the paint is deposited at a thickness inversely proportional to its length. We see that the total thickness is uniform over $[a, b]$, so we conclude that this distribution allows perfect packing. (This argument could easily be formalized.)

We now discuss how the notion of weak closure mentioned above is used by Rhee and Talagrand to exhibit specific densities that allow perfect packing.

Theorem 5.14 ([RT88b]) *Let $[a, b]$ be a subinterval of $[0, 1]$, and let f be a density function defined over this interval such that the corresponding random variable has mean $1/p$ for some integer $p \geq 3$, and such that either*

f *is decreasing over* $[a, b]$ *and* $(p - 1)a + b \leq 1$, *or*

f *is increasing over* $[a, b]$ *and* $a + (p - 1)b \geq 1$.

Then f allows perfect packing.

Corollary 5.15 ([RT88b]) *Throughout the region labeled "Perfect" in Figure 5.2, the uniform distribution over $[a, b]$ allows perfect packing.*

We first illustrate the ideas in the proof of the corollary, and then return to the proof of Theorem 5.14. Consider the region where, for some $p \geq 3$,

$$a \leq \frac{1}{p+1} \leq \frac{a+b}{2} \leq \frac{1}{p} \leq b \leq \frac{1}{2},$$

$$b - a \geq \frac{2}{p(p+1)}, \tag{5.22}$$

and

$$a + (p - 1)b \geq 1.$$

(This and a similar case cover most of the region labeled "Perfect," though a part of that region requires a more difficult proof, which can be found in [RT88b].) For any $[a, b]$ satisfying these inequalities, we decompose $U(a, b)$ into two parts, each of which allows perfect packing by Theorem 5.14, as follows. (See Figure 5.4.) We decompose the density f into two incomplete densities f_1 and f_2, where f_1 is symmetric about $1/(p+1)$ and thus perfectly packable, regardless of our choice of ℓ. Now if $\ell = 0$, the mean of f_2 is at most $1/p$ since $(a + b)/2 \leq 1/p$; on the other hand, if $\ell = (b - a)^{-1}$, the mean of f_2

5.3. FUNCTIONAL ANALYSIS OF THE PACKING CONSTANT

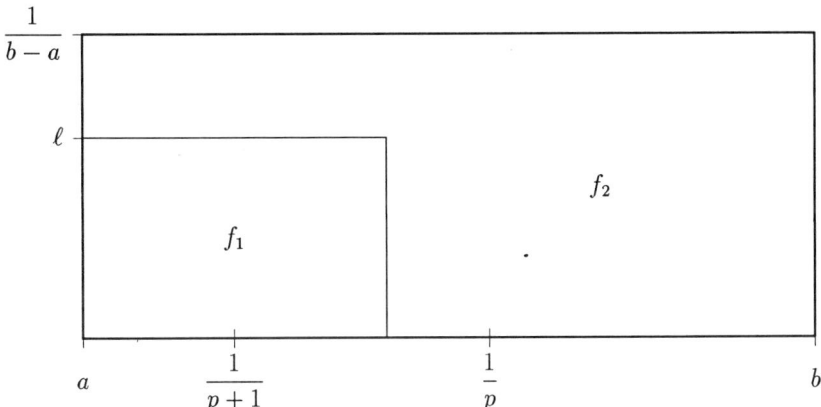

Figure 5.4: A perfect packing strategy for the interval $[a, b]$.

is easily shown from (5.22) to be at least $1/p$. Hence a value $\ell \in [0, (b-a)^{-1}]$ can be chosen so that f_2 has mean exactly $1/p$ and is thus perfectly packable by Theorem 5.14. (Shor [Shor85a] has exhibited explicit packing strategies that give a short alternative proof for much of the region.)

Proof of Theorem 5.14. Fix p. The proof begins by considering the set W of triples (a, b, c) that satisfy $0 \leq a \leq b \leq c \leq 1$, $a + b \leq 2/p$, $c + a \geq 2/p$, and $(p-1)a + c \leq 1$. With each $w = (a, b, c) \in W$, we associate a probability measure μ_w with the following properties:

- If $a \neq b$, μ_w has a constant density on $[a, b)$, and if $b \neq c$, μ_w has a constant density on $[b, c)$,

- μ_w vanishes outside $[a, c)$, and

- μ_w has mean $1/p$.

(If $a < b < c$, then μ_w has a density, but if $a = b$, then μ_w has an atom at a.)

Rhee and Talagrand show that each μ_w allows a perfect packing. The following lemma gives a decomposition crucial to the development; its proof is a tedious case analysis making extensive use of Corollary 5.13, and can be found in [RT88b].

Lemma 5.16 *For each $w \in W$ we can find three nonnegative numbers η_1, η_2, η_3 with $\eta_1 + \eta_2 + \eta_3 < 1$ and three triples w_1, w_2, w_3 such that $\nu = \mu_w - \eta_1\mu_{w_1} + \eta_2\mu_{w_2} + \eta_3\mu_{w_3}$ is a positive measure and allows perfect packing. Moreover, the choices of η_i and w_i are Borel-measurable functions of w.*

Continuation of the proof of Theorem 5.14. Lemma 5.16 suggests that we might proceed as follows to pack μ_w: Decompose μ_w as indicated, and pack ν perfectly; what remains is a positive linear combination of three μ_{w_i}. We can apply the process recursively to each of these. This does not immediately yield a proof, however, since this process is not guaranteed to stop, and it may not be apparent that it even converges in any appropriate sense.

To overcome this problem, let M_W be the set of probability measures over W and begin by noting that any probability measure $\nu \in M_W$ induces a probability measure $\bar{\nu}$ on $[0, 1]$ by

$$\bar{\nu}([0, x]) = \int_W \mu_w([0, x]) \, d\nu(w).$$

(Informally, this means we can generate an x drawn according to $\bar{\nu}$ by first drawing a $w \in W$ according to ν, and then drawing an x according to μ_w.) As we will show in Lemma 5.17 below, for each $\nu \in M_W$, $\bar{\nu}$ allows perfect packing.

It is not hard to demonstrate that any probability measure μ having a density meeting the hypotheses of this theorem can be approximated well by linear combinations of μ_w; more formally, we can find a sequence $\nu_N \in M_W$ such that $\bar{\nu}_N$ converges weakly to μ as $N \to \infty$. Now each $\bar{\nu}_N$ is perfectly packable by Lemma 5.17 below. Hence by Theorem 5.11 μ is perfectly packable. ∎

Lemma 5.17 *For each $\nu \in M_W$, $\bar{\nu}$ allows perfect packing.*

Proof. The proof is analogous to a proof of the simpler fact that, if S is a closed set of nonnegative reals such that for all $x > 0$ in S there exists a $y < x$ in S, then 0 must be in S. Lemma 5.16 implies that for any measure $\nu \in M_W$, there is a probability measure $\nu_1 \in M_W$ and $\eta < 1$ such that $\theta = \bar{\nu} - \eta\bar{\nu}_1$ is a positive measure that allows perfect packing. Letting $\mathcal{W} = \{\bar{\nu} : \nu \in M_W\}$, this means that $\mathcal{W}$ has the property that for any member of $\mathcal{W}$ we can shave off a part of the probability measure that is proportional to some element of $\mathcal{W}$, and have the remainder be perfectly packable.

Given $\nu \in M_W$, let

$$\eta_0 = \inf\{\eta > 0 : \exists \gamma \in M_W \text{ for which } \bar{\nu} - \eta\bar{\gamma} \text{ is a positive measure that allows perfect packing}\}. \tag{5.23}$$

In defining η_0 in this way, we are trying to shave off a part of $\bar{\nu}$ that is proportional to some $\bar{\gamma} \in \mathcal{W}$, so that the mass of what we shave off is as

5.3. FUNCTIONAL ANALYSIS OF THE PACKING CONSTANT

small as possible subject to the remainder being perfectly packable; η is the mass of the part of the probability measure that we are shaving off. By (5.23) there will be sequences η_n, and γ_n, $n > 0$, such that $\lim_{n \to \infty} \eta_n = \eta_0$ and each $\theta_n = \bar{\nu} - \eta_n \bar{\gamma}_n$ is a positive measure allowing perfect packing. Now by compactness γ_n has a weakly converging subsequence, so we may pick a subsequence γ_{n_k} that converges weakly to some $\gamma_0 \in M_W$. Then θ_{n_k} converges weakly to $\bar{\nu} - \eta_0 \bar{\gamma}_0$, and by the weak closure of $\mathcal{C}$ we know that this is perfectly packable. Thus we have used the closure property to guarantee that the infimum above can be replaced by a true minimum; if $\eta_0 = 0$, we are done.

If $\eta_0 \neq 0$, we derive a contradiction as follows. Since $\gamma_0 \in M_W$, we can find $\gamma_0' \in M_W$ and $\eta_0' < 1$ such that $\bar{\gamma}_0 - \eta_0' \bar{\gamma}_0'$ is a positive measure allowing perfect packing. Now note that

$$\bar{\nu} - \eta_0 \bar{\gamma}_0 + \eta_0(\bar{\gamma}_0 - \eta_0' \bar{\gamma}_0') = \bar{\nu} - \eta_0 \eta_0' \bar{\gamma}_0'.$$

Since the left side is a positive linear combination of two perfectly packable measures, it is perfectly packable. But then the right side contradicts (5.23), since we have now shaved off only a mass of $\eta_0' \eta_0 < \eta_0$ to obtain a perfectly packable probability measure. ∎

We conclude this section with a few comments on the work of Courcoubetis and Weber [CW86b]. They proved results for discrete distributions that are analogous to those of Lueker and of Rhee and Talagrand; they also investigated on-line packings in the discrete case. Recall our notation in Section 2.6, where an integer program was defined for bin packing with finite sets of item sizes. Assume here that the item sizes $s_1, \ldots, s_N$ are rationals in $(0, 1]$ and define *perfect* configurations $\vec{c} = (c_1, \ldots, c_N)$ as those for which $\sum_{i=1}^{N} c_i s_i = 1$ and $c_i \geq 0$, $1 \leq i \leq N$. Let the item-size distribution be given by $p_i = \Pr\{X = s_i\}$, $1 \leq i \leq N$. Let $\mathcal{C}^*$ denote the (possibly empty) set of perfect configurations defined on $\{s_1, \ldots, s_N\}$, and consider the convex cone $\Lambda \subseteq \Re^k$ spanned by nonnegative linear combinations of the configurations (vectors) in $\mathcal{C}^*$. It is shown in [CW86b] that the expected number of partially full bins in an optimal packing of n items is $\Theta(n)$, $\Theta(\sqrt{n})$, or $O(1)$ according as the vector $\vec{p} = (p_1, \ldots, p_N)$ lies outside Λ, on the boundary of Λ, or inside Λ, respectively. Clearly, since there are finitely many item sizes and they are strictly positive, the term "partially full bins" can be replaced by "wasted space" in the above assertion. An on-line algorithm is exhibited in [CW86b] that achieves $O(1)$ wasted space in the case where $\vec{p}$ is inside Λ. However, the algorithm is far from practical: it is randomizing, it packs items in a manner that depends on how items are packed in the current set of partially full bins, and it even starts new bins when not required to do so. Subsequent papers have introduced simpler, bounded-waste, on-line algorithms, but there is still no known "simple" algorithm. For example, there are $\vec{p}$ inside Λ for which Best Fit produces $\Theta(n)$ expected wasted space. Other papers have generalized these results by allowing items to be encountered in sequences with correlation or in batches. For further discussion, including descriptions of special cases in which $\vec{p}$ is inside Λ, see [CW86a, CKWW89, CW90].

Chapter 6
Bin Packing: Heuristics

In this chapter we first study off-line algorithms that pack items in decreasing order of size. In Section 6.2 we turn to on-line approaches and cover several topics. The first is a bound on the open-end BF heuristic; recall that BF is an open-end heuristic because it makes no use of the number of items to be packed. A superior closed-end heuristic is then described. We conclude that section with a lower bound on open-end on-line algorithms. In the final section, Section 6.3, we study linear-time algorithms, including one that is asymptotically optimal, and prove a lower bound for heuristics that maintain a limited number of active bins.

Exact results in this chapter are limited chiefly to the more easily analyzed linear-time algorithms. An exact analysis of even the first bin in an FF or FFD packing seems to be quite difficult in general, in spite of the property that the sizes of the items in the first bin are not conditioned on not having fit into earlier bins. This *first-bin* problem for FF, along with its relation to problems of sequential selection, is briefly discussed in [CFW87]; see also [FV89]. Interestingly, when $F(x)$ is the the uniform distribution $U(0,1)$, the problem of determining the number of items packed is easily shown to be equivalent to a classical version of the *record-breaking* problem: in a given sequence $X_1, X_2, \ldots, X_n$ of independent samples from $U(0,1)$, how many times are records set (i.e., for how many indices $j \geq 1$ does $X_j > \max_{1 \leq i < j} X_i$ hold)? Thus from the results on the record-breaking problem (see [Glic78]) we have that, as $n \to \infty$, the number of items packed in the first FF bin is asymptotically normally distributed with mean and variance $\ln n$.

An asymptotic analysis of the first-bin problem for FFD can be found in [CFJR90] (see also [BT89]). This analysis includes a proof that, in contrast to FF, the expected number of items in the first FFD bin is bounded for all n. (In the limit as $n \to \infty$, this expectation is slightly less than $5/3$.)

6.1 Off-line packing: FFD and BFD

6.1.1 The expected behavior

Consider the analysis of the FFD and BFD heuristics described in Section 1.4.2. A technique used in [Fred80, Luek82], at the suggestion of Johnson [John], is an illustration of the method of dominating algorithms (Section 2.4). Here we present a simplified proof, still using dominating algorithms, which benefits from ideas in [BJLMM84, John, Karp82]. Average-case results for FFD and BFD derive quite easily from the following combinatorial result, together with the average-case analysis of MATCH in Chapter 5.

Lemma 6.1 *For any set of item sizes drawn from $[0,1]$, the number of bins used by the heuristic* MATCH *(Section 5.1) is at least as great as the number of bins used by* BFD *or* FFD.

Proof. Consider the following two-stage algorithm for packing a list L_n. In the first stage, a given heuristic H packs the first i items of L_n, for some given $i \geq 0$. The second stage packs the remaining items optimally subject to the constraint that none of these remaining items can be placed into a bin that has already received two or more items. We call the second stage an *Optimum Restricted Completion* (ORC). An easy inductive proof shows that, if we process the items in order of decreasing size, any algorithm of the following form will yield an ORC. Let X be the next item to be packed. If X does not fit into a bin already containing exactly one item, then X is packed into a new bin. Otherwise, X is packed into some bin containing exactly one item. (If more than one such bin exists, the algorithm may pack X into any such bin; this is because items packed later are at most as large as X, so they also fit into bins where X fits.)

Let H_i denote the above two-stage algorithm, with FFD or BFD being the heuristic for packing the first i items of L_n; in either case, an ORC packs the remaining items. Then, in particular, $H_n(L_n)$ denotes either FFD(L_n) or BFD(L_n). Since MATCH gives a packing with at most two items per bin, it must be that $H_0(L_n) \leq \text{MATCH}(L_n)$.[1] We complete the proof by showing that $H_i(L_n)$ is nonincreasing in i, $0 \leq i \leq n$.

For notational convenience assume $X_1 \geq X_2 \geq \cdots \geq X_n$. Suppose first that H_i packs X_i into a bin with at most one item. By the definition of FFD and BFD, X_i must be packed into a bin with another item, if that is

---
[1]In fact, one easily sees that $H_0(L_n) = \text{MATCH}(L_n)$, though we do not need equality for this proof.

6.1. OFF-LINE PACKING: FFD AND BFD

possible. Thus, there is an ORC for H_{i-1} that begins by packing X_i just as X_i is packed by H_i, so $H_i(n) = H_{i-1}(n)$.

Suppose next that H_i places X_i into a bin already containing two or more items. Then that bin is unusable in an ORC for H_{i-1} as well as in an ORC for H_i, i.e., ORCs for H_{i-1} and H_i begin with the same set of bins that have exactly one item. But one readily sees that ORCs are monotonic in the sense that, if the ORCs for H_{i-1} and H_i begin with the same set of usable bins, then the ORC for H_i starts at most as many new bins as the ORC for H_{i-1}. $H_i \leq H_{i-1}$ follows easily from this property and the fact that H_i uses the same number of bins in packing $X_1, \ldots, X_i$ as H_{i-1} uses in packing $X_1, \ldots, X_{i-1}$. ∎

Combined with Theorems 5.1 (page 101) and 5.2 (page 101), Lemma 6.1 proves:

Theorem 6.2 *For item sizes drawn independently from $U(0,1)$, we have* $\mathsf{E}[\text{FFD}(L_n)] - n/2 = \Theta(\sqrt{n})$, *and* $\mathsf{E}[\text{BFD}(L_n)] - n/2 = \Theta(\sqrt{n})$.

Since FFD and BFD use small items to fill gaps left by larger items, it is natural to expect a substantial drop in the average wasted space if the uniform distribution is assumed to extend only over $[0, a]$ for some $a < 1$. Indeed, Bentley et al. [BJLMM84] showed that expected wasted space under FFD with $a \leq 1/2$ is bounded by a constant independent of n. The proof amounted to an analysis of order statistics and their relation to the structure of FFD packings. The analysis is too involved to present here, but we can illustrate informally the nature of their approach. Define a *regular* item in an FFD packing as an item that, at its time of packing, either starts a new bin or is placed in the highest indexed (i.e., last) nonempty bin. The remaining items are called *fallback* items for obvious reasons. Bentley et al. showed that if the gap left over by the regular items in a bin B_i falls in the interval between the jth and $(j + 1)$th order statistics, $X_{(j)}$ and $X_{(j+1)}$, then with high probability for all n sufficiently large, the first fallback item packed in B_i is an order statistic $X_{(l)}$, where $j - l$ is bounded by a constant independent of n. Thus we can expect B_i to be nearly full.

Among other results, [BJLMM84] showed that the average wasted space as a function of n exhibits unexpected discontinuities at $a = 1/2$ and $a = 1$.

Theorem 6.3 ([BJLMM84]) *Let $U(0, a)$ be the item-size distribution. Then*[2]

$$\mathsf{E}[\text{FFD}(L_n) - \sum_{i=1}^{n} X_i] = \begin{cases} \Theta(1) & \text{if } a \leq 1/2, \\ \Theta(n^{1/3}) & \text{if } 1/2 < a < 1. \end{cases}$$

---
[2] The $\Theta(n^{1/3})$ result appears in a paper in preparation, which is based on [BJLMM84] and carries the same authors.

(Recall from Theorem 6.2 that the average wasted space under FFD is $\Theta(n^{1/2})$ when $a = 1$.)

Floyd and Karp [FK] have studied FFD, $a = 1/2$, under the assumption that the number of items is a Poisson random variable N with $\mathsf{E}[N] = n$ (see Section 2.7.1). Thus the distances between adjacent item sizes are i.i.d. exponential random variables with mean values $1/n$. The analysis in [FK] is based on a modification of FFD, called MFFD, that closes a bin once it receives a fallback item. It is not difficult to prove that MFFD dominates FFD. Evaluating the performance of MFFD involves an ingenious application of queueing theory that yields two insights:

a) a simple, heuristic explanation for the discontinuity at $a = 1/2$ in Theorem 6.3, and

b) a constant upper bound to the expected wasted space that is much smaller than that implied by the analysis of [BJLMM84].

Before getting into the analysis, we first show how the behavior of MFFD can be represented by a queueing process.

We describe MFFD in terms of a right-to-left scan of the interval $[0, 1/2]$ pictured in Figure 6.1, on which a sample, L_N, of item sizes has been plotted as a pattern of dots. MFFD uses a FIFO queue, Q, that is initially empty; later in the scan, Q contains the sizes of gaps in partially packed bins that have yet to receive fallback items (the scan has yet to encounter items small enough to fit into the gaps represented in Q).

Suppose that at some point in the scan, MFFD encounters a dot at x. If Q is empty, then the corresponding item is packed into the highest indexed, nonempty bin, say, B_j, if it fits. If it does not fit, then it starts the new bin, B_{j+1}, and the current gap g_j in B_j is plotted as a cross at point $g_j < x$. If Q is not empty when MFFD encounters the dot at x, then the item is placed into the bin whose gap $g_i > x$, $i < j$, appears at the head of Q, whereupon g_i is removed from Q. Finally, whenever MFFD encounters a cross, the corresponding gap size is simply appended to Q.

We associate a queueing process with MFFD by envisioning MFFD's leftward scan to be at rate 1, and letting $q(t)$ be the number of customers (gaps) in Q at time t (when the scan is at $\frac{1}{2} - t$). The arrival time of a customer g in Q is simply $\frac{1}{2} - g$, the time at which the corresponding cross is encountered. The sojourn time (or time spent in Q) of g is $g - x$ if an item at $x < g$ is placed into the bin with gap g; the sojourn time is simply g if no further item is ever placed into this bin (see Figure 6.1). The sojourn time of a gap g corresponding to B_i is clearly equal to the gap of unused

6.1. OFF-LINE PACKING: FFD AND BFD

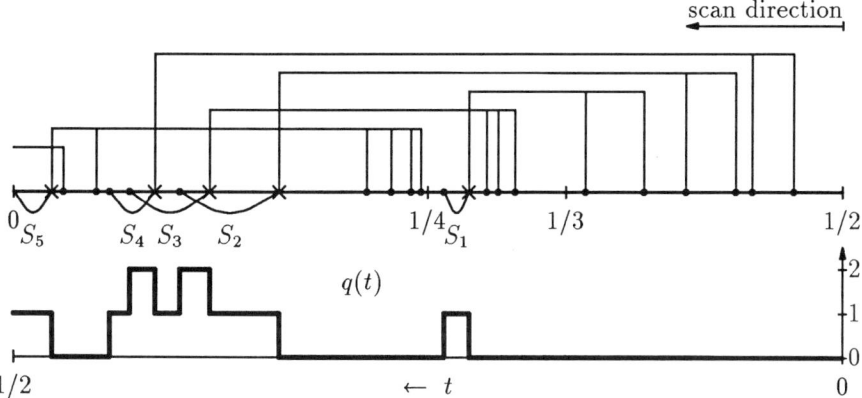

Figure 6.1: Example of the queueing process for MFFD. S_i denote sojourn times. The bin corresponding to a cross contains the matched fallback item (if any) to the left of the cross, and all regular items matched to the cross and to the right of the cross.

space remaining in B_i at the end of the scan (packing process). Thus the cumulative wasted space in the packing up to but not including the last bin is simply the sum of all the sojourn times. Note that our interest in $q(t)$ can be restricted to $1/6 \leq t \leq 1/2$, since no gap exceeding $1/3$ is ever created, i.e., $q(t) = 0$ for all $0 \leq t \leq 1/6$.

The area under the line $q(t)$ in Figure 6.1 can be interpreted in two ways. First, it is the sum of the gaps in all but the last bin at the end of the scan, by the remarks of the previous paragraph. But also, of course, it is simply $1/3$ times the average value of $q(t)$ over $t \in [1/6, 1/2]$. Thus, if the queueing process is stable and hence $\mathsf{E}[q(t)]$ is bounded by some absolute constant, then the expected wasted space is bounded by a constant.

It is helpful to partition $q(t)$ into component processes $\{q_k(t)\}_{k \geq 2}$, where $q_k(t)$ is simply $q(t)$ restricted to the time interval $I_k = [\frac{1}{2} - \frac{1}{k}, \frac{1}{2} - \frac{1}{k+1})$ corresponding to the interval $(\frac{1}{k+1}, \frac{1}{k}]$ of item sizes. To analyze $q_k(t)$, we first consider its arrival process.

Packing bins with k regular items having sizes in $(\frac{1}{k+1}, \frac{1}{k}]$ creates gap sizes in $(0, \frac{1}{k+1}]$; in $q(t)$ these are customers arriving in $[\frac{1}{2} - \frac{1}{k+1}, \frac{1}{2}) = \cup_{j \geq k+1} I_j$. As a convenient approximation, let us assume that these arrivals form a Poisson pattern over the intervals I_j, $j \geq k+1$, with a density determined as follows. Let η_k be the density of regular items with sizes in $(\frac{1}{k+1}, \frac{1}{k}]$. We have $\eta_2 = n$, but because of fallback items, $\eta_k < n$ for $k > 2$. Then the packing of items with sizes in $(\frac{1}{k+1}, \frac{1}{k}]$ creates an expected number of customers in $\cup_{j \geq k+1} I_j$

that is asymptotic to $(1/k)(\eta_k/k(k+1))$ as $n \to \infty$. Since $\cup_{j \geq k+1} I_j$ has duration $1/(k+1)$, we take η_k/k^2, $k \geq 2$, as the arrival rate of the Poisson process generated by the scan of item sizes in $(\frac{1}{k+1}, \frac{1}{k}]$.

To determine the η_k, observe first that the total arrival rate λ_{k+1} of customers to I_{k+1} originating from the scan of the item sizes in $(\frac{1}{3}, \frac{1}{2}], \ldots, (\frac{1}{k+1}, \frac{1}{k}]$ is given by

$$\lambda_{k+1} = \sum_{j=2}^{k} \eta_j/j^2 \quad \text{for } k \geq 2. \tag{6.1}$$

Now $\eta_{k+1} = n - \lambda_{k+1}$, so we determine η_k and λ_k from the recurrence

$$\eta_2 = n,$$

$$\eta_{k+1} = n - \sum_{j=2}^{k} \eta_j/j^2,$$

for which the solution is easily found to be

$$\eta_k = \frac{n}{2} \frac{k}{k-1} \quad \text{for } k \geq 2. \tag{6.2}$$

Substituting into (6.1), we obtain the arrival rates for the processes $q_k(t)$,

$$\lambda_{k+1} = \frac{n}{2} \sum_{j=2}^{k} \frac{1}{j(j-1)} = \frac{n}{2} \frac{k-1}{k} \quad \text{for } k \geq 2, \tag{6.3}$$

with $\lambda_2 = 0$.

We observe immediately from (6.3) that $\lambda_k < n/2$ for all $k \geq 2$. These arrival rates are all exceeded by the maximum service rate n, so our Poisson approximation implies that $q_k(t)$ is stable for all $k \geq 2$, and hence that $q(t)$ is stable. This provides a strong basis for expecting that

$$\mathsf{E}\left[\mathrm{FFD}(L_n) - \sum_{i=1}^{n} X_i\right] \leq \mathsf{E}\left[\mathrm{MFFD}(L_n) - \sum_{i=1}^{n} X_i\right] = O(1).$$

But notice what happens when we extend the model in the obvious way to items at density n over $(0, a]$ for $a > 1/2$. Since the items in $[1/2, a)$ go one to a bin, the arrival rate of customer gaps throughout $[1-a, 1/2)$ is at least the maximum service rate n. Thus $q(t)$ is unstable in $[1-a, 1/2)$ and $\mathsf{E}[\mathrm{MFFD}(L_N) - \sum_{i=1}^{n} X_i]$ is unbounded as $n \to \infty$. The experiments in [BJLM83] give convincing evidence that $\mathsf{E}[\mathrm{MFFD}(L_N)] - \mathsf{E}[\mathrm{FFD}(L_N)] = O(1)$, so the above analysis leads us to expect the discontinuity at $a = 1/2$ in Theorem 6.3.

6.1. OFF-LINE PACKING: FFD AND BFD

The experiments of [BJLM83] also suggest that the asymptotic expected wasted space is very close to 0.7 as $n \to \infty$. By a natural extension of our Poisson approximation, a simple queueing analysis yields a result remarkably close to this. (This approximation is attributed to Coffman by Floyd and Karp.) Specifically, let us take the sample functions of $q_k(t)$ as those of a single-server queue in statistical equilibrium with Poisson arrivals at rate λ_k and i.i.d. service times exponentially distributed with mean $1/n$. For the expected total sojourn time of customers in I_{k+1}, $k \geq 2$, we have

$$W_{k+1} \sim N_{k+1} w_{k+1} \quad \text{as} \quad n \to \infty, \tag{6.4}$$

where $N_{k+1} = \lambda_{k+1}/(k+1)(k+2)$ is the expected number of customers in I_{k+1}, and where

$$w_{k+1} = \frac{1/n}{1 - \lambda_{k+1}/n}$$

is the expected waiting time of each such customer (see [Klei76, p. 98]). Substitution into (6.4) yields, for $k \geq 2$,

$$W_{k+1} \sim \frac{k-1}{2k(k+1)(k+2)} \cdot \frac{1}{1-(k-1)/2k} = \frac{k-1}{(k+1)^2(k+2)} \quad \text{as} \quad n \to \infty.$$

Summing over $k \geq 2$ and adding 0.5 for the approximate expected wasted space in the last bin leads to the approximation

$$\mathsf{E}[\mathrm{MFFD}(L_N)] - \sum_{i=1}^{N} X_i = 0.5 + \sum_{k \geq 2} \frac{k-1}{(k+1)^2(k+2)} \approx 0.71.$$

A rigorous analysis of the Floyd-Karp model faces serious difficulties. Our Poisson approximation fails because

a) the $q_k(t)$, $k \geq 2$, are not independent processes, and
b) the interarrival times of any given $q_k(t)$, $k > 2$, are not mutually independent.

Floyd and Karp eliminate the first problem by introducing a further, still dominating modification of MFFD. Even after this modification, the second difficulty remains. Resorting to bounding techniques, they prove

Theorem 6.4 ([FK]) *With L_N given by the renewal points in $[0, 1/2]$ of a Poisson process at rate $n = \mathsf{E}[N]$, we have $\mathsf{E}[\mathrm{FFD}(L_N)] - \sum_{i=1}^{N} X_i \leq 9.4$.*

Although this upper bound appears to be crude, as we have seen, we emphasize that it is a rigorous bound many orders of magnitude smaller than the bound implied by the analysis in [BJLMM84].

6.1.2 Deviation from the expected behavior

Given two lists L_n and $L'_{n'}$, let the corresponding sample distribution functions[3] be F_n and $F'_{n'}$. As noted in Section 2.6, neither BFD nor FFD is monotonic. In fact, Murgolo [Murg88] gives examples of lists L_n and $L'_{n'}$, with $n = n'$ and L_n dominating $L'_{n'}$, for which

$$\text{BFD}(L'_n) \geq \frac{43}{42} \text{BFD}(L_n).$$

In terms of the sample distribution functions, this means that we can simultaneously have

$$\forall t, \ F'(t) \geq F(t) \ \text{ and } \ \text{BFD}(L'_n) = \bigl(1 + \Omega(1)\bigr) \text{BFD}(L_n).$$

Rhee and Talagrand [RT89b] establish, however, that by bounding the difference between $F'(t)$ and $F(t)$ (so that $F'(t)$ can be neither much more than nor much less than $F(t)$), we can cause $\text{BFD}(L'_n)$ to be arbitrarily close to $\text{BFD}(L_n)$. Thus BFD is continuous in some sense. More precisely, let the *sample distance* between L_n and $L'_{n'}$ be $\sup_{0 \leq t \leq 1} |F_n(t) - F'_{n'}(t)|$. Then there is a function $\phi(\delta, n)$ such that

$$\lim_{\delta \to 0} \lim_{n \to \infty} \phi(\delta, n) = 0,$$

and

$$\left| \frac{\text{BFD}(L_n)}{n} - \frac{\text{BFD}(L'_{n'})}{n'} \right| \leq \phi(\delta, n),$$

where δ is the sample distance between L_n and L'_n.

For FFD, which appears to be even more pathological than BFD (see [RT89b, page 910]), a weaker continuity result is proved in [RT89c]: for each distribution F, there exists a constant c such that for each $\epsilon > 0$, there are δ and n_0 such that

$$n \geq n_0 \text{ and } \sup_{0 \leq t \leq 1} |F_n(t) - F(t)| \leq \delta \implies \left| \frac{\text{FFD}(L_n)}{n} - c \right| \leq \epsilon.$$

With these bounds established, Rhee and Talagrand can easily obtain the following convergence result from properties of the Kolmogorov-Smirnov statistic:

---
[3]See Section 2.7.2 (page 34) for the definition of the term sample distribution function.

Theorem 6.5 ([RT89b, RT89c]) *Let H be one of BFD or FFD, and let F be a probability distribution. As usual let the list L_n consist of n i.i.d. random variables with distribution F. Then there exists a constant c such that for every $\epsilon > 0$,*

$$\sum_{n=1}^{\infty} \Pr\left\{ \left| \frac{H(L_n)}{n} - c \right| \geq \epsilon \right\} < \infty.$$

6.2 On-line bin packing: Best Fit

Shor [Shor86] discovered the intimate connection between up-right matching (Section 3.2) and the structure of BF packings. In so doing, he produced a remarkably precise estimate of the expected wasted space. The analysis below follows that in [Shor86].

To prove an upper bound, the dominating-algorithm technique will again supply a major simplification. Define *Modified Best Fit* (MBF) to be the same as BF except that a bin is closed whenever it receives an item no larger than 1/2. Thus bins in an MBF packing have at most two items. Also, at any point in the packing sequence, the active bins are just those singleton bins with items exceeding 1/2. The dominance of MBF is proved below.

Lemma 6.6 *For all lists L_n, $\mathrm{MBF}(L_n) \geq \mathrm{BF}(L_n)$.*

Proof. The proof exploits the fact that MBF, in contrast to BF, is monotonic. Indeed, it is easy to prove by induction that the following more descriptive property holds. Let $L_n^{(i)}$ denote the list L_n with the ith item X_i deleted, $1 \leq i \leq n$. Then, in terms of the number of one-item and two-item bins, the MBF packings of L_n and $L_n^{(i)}$ can differ in only two ways: the MBF packing of L_n either has

i) one fewer one-item and one more two-item bin, or

ii) the same number of two-item bins and one more one-item bin.

This property implies that $\mathrm{MBF}(L_n) \geq \mathrm{MBF}(L_n^{(i)}) \geq \mathrm{MBF}(L_n) - 1$.

Now remove from any given list L_n all items that are packed by BF into a bin already containing an item no larger than 1/2. Then the new list L_n' contains no item that is packed by BF into a bin that MBF would consider closed. Thus MBF and BF pack the items in L_n' in exactly the same way. By the monotonicity of MBF, we then have

$$\mathrm{MBF}(L_n) \geq \mathrm{MBF}(L_n') \geq \mathrm{BF}(L_n). \qquad \blacksquare$$

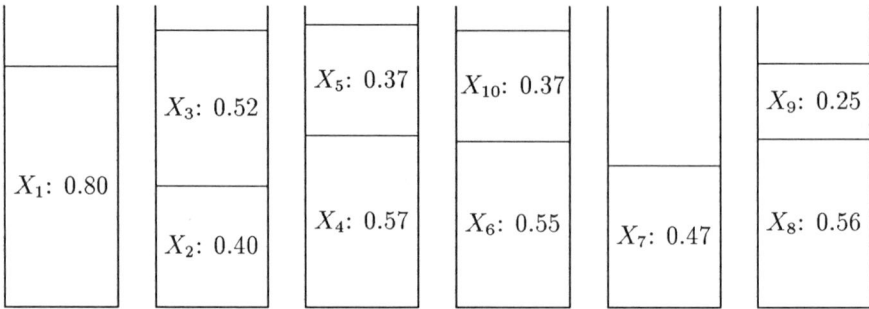

Figure 6.2: An example of Best Fit.

Theorem 6.7 ([Shor86]) *For n i.i.d. items from $U(0,1)$, we have the expected wasted space*

$$\mathsf{E}[\mathrm{BF}(L_n)] - \frac{n}{2} = \Theta(\sqrt{n}\log^{3/4} n).$$

Proof of the upper bound. By Lemma 6.6 we need only prove the $O(\sqrt{n}\log^{3/4} n)$ upper bound for MBF. Let the items of L_n be points plotted on a square as follows: the vertical coordinate corresponds to the index of the item; larger indices appear lower in the square. The horizontal coordinate is just the size of the item. We plot a $\oplus$ for items greater than $1/2$, and a $\ominus$ for items less than $1/2$. Next we fold the square in half vertically; equivalently, we can think of plotting points X_i that are greater than $1/2$ at $1 - X_i$. We now match points that are placed into the same bin by MBF. (See Figure 6.2 for an example of a set of items to be packed and their packing according to Best Fit; see Figure 6.3 for an example of the up-right matching for the data of Figure 6.2.)

Note that MBF always packs an item of size greater than $1/2$ into a new bin, and packs an item of size less than $1/2$ into the bin containing a single item of size at least $1/2$ and having the smallest gap no smaller than the item, provided such a bin exists. In terms of the points we have plotted, this means that the corresponding matching can be produced by scanning the $\ominus$ points from top to bottom, matching each $\ominus$ with the leftmost unmatched $\oplus$ that is above it (has a lower index) and to its right (and, therefore, will fit together with it in a bin). It is easily proved that this always produces a maximum up-right matching [KLMS84]. But, as we saw in Theorem 3.2 (page 43), the expected number of unmatched points in a maximum up-right matching is $O(\sqrt{n}\log^{3/4} n)$. ∎

6.2. ON-LINE BIN PACKING: BEST FIT

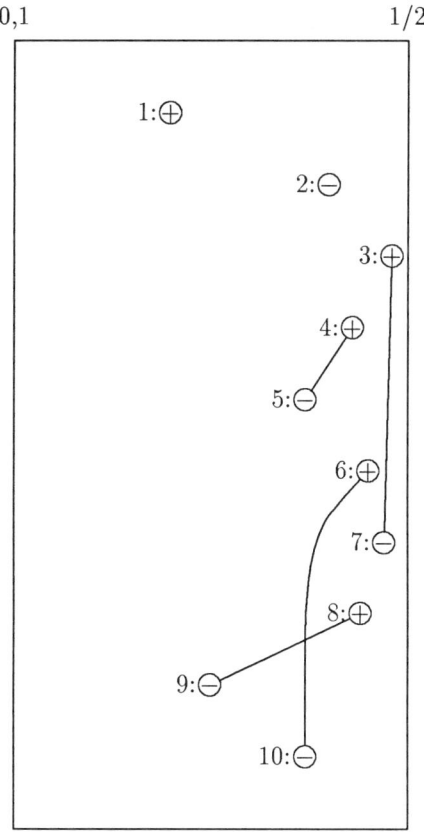

Figure 6.3: An example of the matching used during the proof of the upper bound on Best Fit. Note that items 3 and 7 are matched even though Best Fit does not pack them together.

We limit our discussion of the lower bound to a couple of hints (for details, see [Shor86]). First it is convenient to restrict the analysis to that part of the packing containing items in the range $(1/3, 2/3)$, which are packed at most two per bin. This part of the packing by itself implies the $\Omega(\sqrt{n}\log^{3/4} n)$ lower bound. The objective is then to show that sufficiently many bins are packed by first putting in a large item $X > 1/2$ and then putting in a small item $X \leq 1/2$. Such pairs give an up-right matching that has an expected value of $\Omega(\sqrt{n}\log^{3/4} n)$ unmatched points, by Theorem 3.2. The expected number of unmatched points will be within a constant factor of the expected number of bins in excess of $n/2$.

Shor also establishes that First Fit wastes an amount of space that is $\Omega(n^{2/3})$ and $O(n^{2/3}\log^{1/2} n)$ when packing n items drawn uniformly from $[0, 1]$. The upper bound again uses the notion of two-dimensional matching. Quite recently, the upper bound has been tightened to give $\mathsf{E}[\mathrm{FF}(L_n) - n/2] = \Theta(n^{2/3})$; matching results continue to play the key role [CCGJM&91].

Surprisingly, there is an on-line packing algorithm that achieves only $\Theta(\sqrt{n})$ wasted space [Shor86]: pack the first half of the items into separate bins, and then pack the remaining items using Best Fit; note that this simple algorithm is able to do well because it knows in advance the number of items to be packed, i.e., it is closed-end. Shor has shown that an open-end on-line algorithm, such as BF or FF, cannot achieve an $O(\sqrt{n})$ wasted space bound.

Theorem 6.8 ([Shor86]) *Let k be chosen uniformly from $\{1, 2, \ldots, n\}$ and let an open-end on-line algorithm pack k items chosen uniformly from $[0, 1]$. Then the expected amount of wasted space in the packing is $\Omega(\sqrt{n \log n})$.*

Proof. With minor changes, we proceed as in [Shor86] and relate this problem to a two-dimensional matching problem. Suppose that at some point in time, n_1 is the number of items in the range $(1/2, 1]$, and n_2 is the number of items in the range $(1/3, 1/2]$ that are not packed in a bin with an item greater than $1/2$. Then, since a single bin cannot hold more than one item of size greater than $1/2$, nor more than two items of size greater than $1/3$, we conclude that the number of bins in use must be at least $n_1 + n_2/2$; hence, since the expected total size of the items and the expectation n_1 are both equal to $n/2$, n_2 gives us a lower bound on the growth rate of the expected wasted space.

To estimate n_2, note that we need not consider items of size greater than $2/3$ as candidates for matching. We now plot $\oplus$'s and $\ominus$'s as before, folding about $1/2$, but now we concentrate on the range $[1/3, 2/3]$. For each X_i, we

6.3. ON-LINE LINEAR-TIME BIN PACKING

plot

$$\begin{cases} \text{nothing} & \text{if } x_i \in [\,0\,,1/3], \\ \ominus \text{ at } (\frac{i-1}{n}, x_i) & \text{if } x_i \in (1/3\,,1/2], \\ \oplus \text{ at } (\frac{i-1}{n}, 1-x_i) & \text{if } x_i \in (1/2\,,2/3], \\ \text{nothing} & \text{if } x_i \in (2/3\,,\,1\,]. \end{cases}$$

We join two points of opposite sign by a line segment if they correspond to items that were packed into the same bin. In addition, we draw a line segment down to the bottom axis from each point that was never packed together with a point of opposite sign. See Figure 6.4 for an example of this construction for the Best Fit packing of Figure 6.2. Note that a horizontal line segment drawn at any point in time (other than a multiple of $1/n$) will intersect a number of lines equal to the number of unmatched points. Thus the expected number of unmatched points, given that we choose k uniformly from $\{1, 2, \ldots, n\}$, is equal to the expected total vertical length of these segments. By Theorem 3.4, page 54 (see also the remarks at the beginning of Section 3.1.3) this is $\Omega(\sqrt{n \log n})$, from which the result follows. ∎

Very recent results of Shor [Shor90] have yielded an open-end algorithm achieving $O(\sqrt{n \log n})$ wasted space. The algorithm grew out of the recent upper-bound result for rightward matching (cf. the comment at the end of Section 3.1). For an application of up-right matching to on-line packings of items from a general distribution, see [RT91].

6.3 On-line linear-time bin packing

In this section we study the analysis of algorithms in which the time to pack an item is bounded by a constant independent of n. The simplest of these is Next Fit, which we consider first. Next, we discuss briefly the asymptotic analysis of the harmonic algorithms. By sacrificing some of the simplicity of NF, the harmonic algorithms offer substantial improvements in packing efficiency.

6.3.1 Next Fit: The expected behavior

The simplicity of NF accounts for its being analyzed successfully within the classical theory of Markov processes. For the analysis it is convenient to define $L = (X_1, X_2, \ldots)$ as an unbounded sequence of i.i.d. random variables (items) with the distribution $F(x)$ on $[0, 1]$. L_n then denotes the first n items of L.

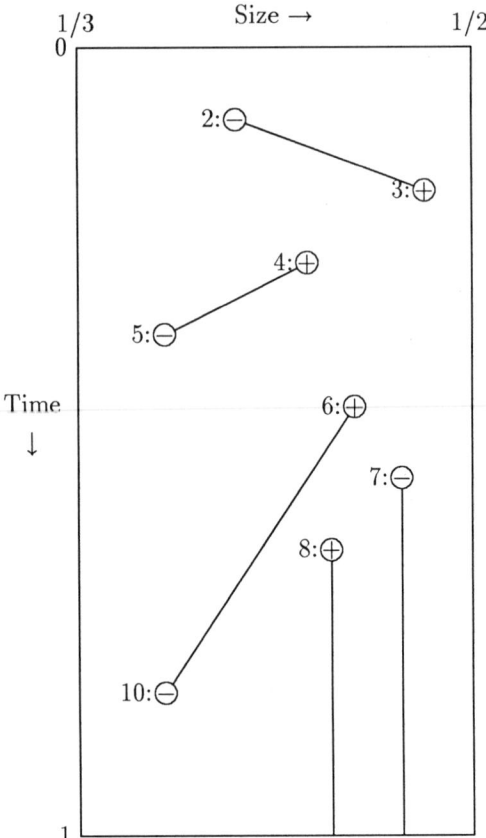

Figure 6.4: An example of the matching used during the proof of the lower bound for open-end on-line packing.

6.3. ON-LINE LINEAR-TIME BIN PACKING

We analyze two Markov chains. First, we analyze $\{l_i\}_{i\geq 1}$, where l_i denotes the final level of B_i, i.e., the level of B_i when B_{i+1} receives its first item. It is easy to verify that $\{l_i\}$ is a first-order Markov chain. Under mild conditions on $F(x)$, the general theory provides that $\{l_i\}$ is ergodic and that $V_i(x) = \Pr\{l_i \leq x\}$ converges geometrically fast to a limiting distribution $V(x)$; see [CSHY80], where it is also verified that there exists a constant $c > 0$ such that for all m

$$\left| m\alpha - \sum_{i=1}^{m} \mathsf{E}[l_i] \right| < c, \tag{6.5}$$

where $\alpha = \lim_{i\to\infty} \mathsf{E}[l_i]$. In particular, this holds if F is uniform over $[0,1]$.

The second Markov chain to be analyzed is $\{m_i\}_{i\geq 1}$, where m_i denotes the level of the last occupied bin (the "current" bin) just after the ith item is packed. As will be seen, asymptotic results on $\mathsf{E}[\mathrm{NF}(L_n)]$ can be obtained for distributions $U(0,a)$, $0 < a \leq 1$, more easily from the analysis of $\{m_i\}_{i\geq 1}$. The convergence properties of $\{l_i\}$ also apply to $\{m_i\}$.

For the analysis of $\{l_i\}$, we define the conditional probability

$$K(x,y) = \Pr\{l_{i+1} \leq y \mid l_i = x\},$$

so that the stationary distribution is defined by

$$V(y) = \int_0^1 K(x,y)\, dV(x). \tag{6.6}$$

Let S_n denote the sum of n i.i.d. samples X from $F(x)$, and let Z_i denote the first item packed into B_i. If $l_i = x$, then Z_{i+1} has the conditional distribution of an item size given that it is larger than $1 - x$, i.e.,

$$\Pr\{Z_{i+1} \leq z \mid l_i = x\} = \begin{cases} 0 & \text{for } z \leq 1 - x, \\ \dfrac{F(z) - F(1-x)}{1 - F(1-x)} & \text{for } 1 - x < z \leq 1. \end{cases} \tag{6.7}$$

Then the probability that exactly j more items fit and the total space used in B_{i+1} is bounded by $y \leq 1$ is $\Pr\{Z_{i+1} + S_j \leq y \text{ and } Z_{i+1} + S_j + X > 1\}$, so

$$K(x,y) = \begin{cases} \displaystyle\sum_{j\geq 0} \Pr\{1 - X < Z_{i+1} + S_j \leq y \mid l_i = x\} & \text{for } y > 1 - x, \\ 0 & \text{for } y \leq 1 - x. \end{cases} \tag{6.8}$$

After computing $V(x)$ and then α, either from an explicit formula or numerically, $\mathsf{E}[\mathrm{NF}(L_n)]$ can be estimated as follows. First, it is easily verified that

$$n\mathsf{E}[X] \sim \mathsf{E}\left[\sum_{i=1}^{\mathrm{NF}(L_n)} l_i\right] \quad \text{as } n \to \infty.$$

Then, assuming that F is such that (6.5) holds, one can prove [CSHY80]

$$\mathsf{E}\left[\sum_{i=1}^{\mathrm{NF}(L_n)} l_i\right] \sim \alpha \mathsf{E}[\mathrm{NF}(L_n)] \quad \text{as } n \to \infty,$$

so that

$$\mathsf{E}[\mathrm{NF}(L_n)] \sim n\mathsf{E}[X]/\alpha \quad \text{as } n \to \infty. \qquad (6.9)$$

In general, explicit solutions for $V(x)$ appear hard to come by. However, for $F(x)$ uniform on $[0,1]$ a simple solution can be found. Under this assumption, $\Pr\{S_j \leq x\} = x^n/n!$, $0 \leq x \leq 1$, and

$$\Pr\{Z_{i+1} \leq z \mid l_i = x\} = \frac{1}{x}(x + z - 1) \quad \text{for } 1 - x < z \leq 1. \qquad (6.10)$$

By (6.8), a calculation then shows that

$$K(x,y) = \begin{cases} 1 - \frac{1}{x}(1-y)e^{x-(1-y)} & \text{for } 1 - x < y \leq 1, \\ 0 & \text{for } y \leq 1 - x. \end{cases} \qquad (6.11)$$

It is easily verified that $V(x)$ has a density $v(x)$. Substitution of (6.11) into (6.6) and differentiation yield

$$v(y) = ye^{y-1} \int_{1-y}^{1} \frac{e^x}{x} v(x)\,dx. \qquad (6.12)$$

In terms of the function $g(y) = e^{-y}v(y)/y$, this integral equation is easily converted to a second-order differential equation with constant coefficients. Ultimately, one obtains $g(y) = 3ye^{-y}$, which proves

Theorem 6.9 ([CSHY80]) *For $F(x)$ uniform over $[0,1]$, we have*

$$V(x) = x^3 \quad \text{for } 0 \leq x \leq 1,$$

and hence

$$\mathsf{E}[\mathrm{NF}(L_n)] \sim \frac{2}{3}n \quad \text{as } n \to \infty.$$

6.3. ON-LINE LINEAR-TIME BIN PACKING

From (6.10) and Theorem 6.9 the limiting density $h(z)$ of Z_i as $i \to \infty$ is easily found. A calculation gives

$$h(z) = \frac{3}{2}(2z - z^2) \quad \text{for } 0 \leq z \leq 1,$$

with the expected value 5/8. The limiting distribution $\{q_m\}$ of the number per bin can then be obtained from

$$q_m = \int_0^1 h(z)\,dz \int_0^{1-z} \Pr\{S_{m-1} = s,\ X > 1 - z - s\}\,ds.$$

Routine manipulations yield

$$q_m = 3\frac{m^2 + 3m + 1}{(m+3)!} \quad \text{for } m \geq 1, \tag{6.13}$$

with the expected value 3/2.

Even for extensions to the distributions $U(0, a)$, $0 < a < 1$, the above method for calculating $V(x)$ runs into major difficulties. However, if we want only the asymptotics of $E[NF(L_n)]$ for these distributions, then the following observation suggests another approach. Suppose we can calculate $\rho = \lim_{i \to \infty} \rho_i$, where ρ_i is the probability that the ith item begins a new bin. Then ρ is the asymptotic fraction of the items starting new bins and

$$E[NF(L_n)] \sim \rho n \quad \text{as } n \to \infty. \tag{6.14}$$

Now let $W(x)$ be the stationary distribution for the Markov chain $\{m_i\}$, where m_i is the level of the current bin after packing the ith item. Then the limiting probability that the next item starts a new bin is

$$\rho = \int_{y=0}^1 \int_{x=1-y}^1 f(x)\,dx\,dW(y),$$

so for the uniform distribution on $[0, a]$, $f(x) = 1/a$ for $0 \leq x \leq a$, and an integration yields

$$\rho = 1 - \frac{1}{a}\int_{1-a}^1 W(y)\,dy. \tag{6.15}$$

Following the analysis of Karmarkar [Karm82] (see also [Hofr87][4]), we now show how to calculate $W(x)$ and hence ρ for the case $f(x) = 1/a$, $0 \leq x \leq a$. First, it is easily verified that $W(x)$ has a density $w(x)$ satisfying

$$w(x) = \int_0^x f(x-y)w(y)\,dy + \int_{1-x}^1 f(x)w(y)\,dy. \tag{6.16}$$

---
[4] We have incorporated Hofri's corrections to [Karm82].

Note that the first term corresponds to fitting an item of size $x - y$ into a bin with level y, and the second term corresponds to starting a new bin with an item of size x because it did not fit into a bin with level $y > 1 - x$. Substituting for $f(\cdot)$ and putting the result in terms of the distribution $W(x)$, we obtain

$$a\frac{dW(x)}{dx} = \begin{cases} W(x) - W(1-x) + 1 & \text{for } x \leq a, \\ W(x) - W(x-a) & \text{for } x > a. \end{cases} \quad (6.17)$$

In what follows, (6.17) is solved by reducing it to two systems of linear first-order differential equations with constant coefficients. Although the details differ from those in [Karm82], the general method is the same. This method is also nicely illustrated in [CCF84].

We assume initially that $1/(K+1) < a < 1/K$ for some $K \geq 1$. Later, we consider the simplifications that occur when $a = 1/K$, $K \geq 1$. The first step is to partition the intervals $[0, a]$ and $[a, 1]$ into a smallest set of subintervals such that for $K \geq 2$ a subinterval of $[0, a]$ maps onto a subinterval of $[a, 1]$ under the reflection $R(x) = 1 - x$ in the first case of (6.17), and a subinterval of $[a, 1]$ maps onto a subinterval of either $[0, a]$ or $[a, 1]$ under the translation $T(x) = x - a$ in the second case. An alternating merge of the two sequences $[ja, 1 - (K - j)a]$, $0 \leq j \leq K$, and $[1 - (K - j)a, (j + 1)a]$, $0 \leq j \leq K - 1$, yields the desired partition of $[0, 1]$, as illustrated in Figure 6.5. The labeled arrows in Figure 6.5 indicate the mappings created by (6.17). The intervals of the form $[ja, 1 - (K - j)a]$ are called *type-1 intervals* and have length $1 - Ka$; the others are called *type-2 intervals* and have length $(K + 1)a - 1$. For $K = 1$, Figure 6.5 shows the special case of one type-2 interval, $[1 - a, a]$, which maps into itself under the reflection $R(x)$.

Now define

$$h(x) = \begin{cases} R(x) & \text{for } 0 \leq x \leq a, \\ T(x) & \text{for } a \leq x \leq 1, \end{cases} \quad (6.18)$$

and let $h^i(x) = h\big(h^{i-1}(x)\big)$, $i \geq 1$, where $h^0(x) = x$. It is clear from (6.17) that $W(x)$ depends on $W\big(h(x)\big)$ which depends on $W\big(h^2(x)\big)$ which depends on $W\big(h^3(x)\big)$, etc.

The second step is to verify that the sequence $h^i(x)$, $i \geq 0$, is periodic. But it is easily seen that if x is in a type-1 interval, then $h^{2(K+1)}(x) = x$, and if x is in a type-2 interval, then $h^{2K}(x) = x$. For example, consider $x \in [0, 1 - Ka]$ and the sequence of type-1 intervals. Figure 6.6 shows the sequence $h^i(x)$, $0 \leq i \leq 2K + 2$, defined by (6.18). Observe that $h(x), h^2(x), \ldots, h^{K+1}(x)$ are $R(x), T(R(x)), \ldots, T^K(R(x))$, respectively; this sequence maps the point x

6.3. ON-LINE LINEAR-TIME BIN PACKING

a) $K = 1$, $\frac{1}{2} < a < 1$

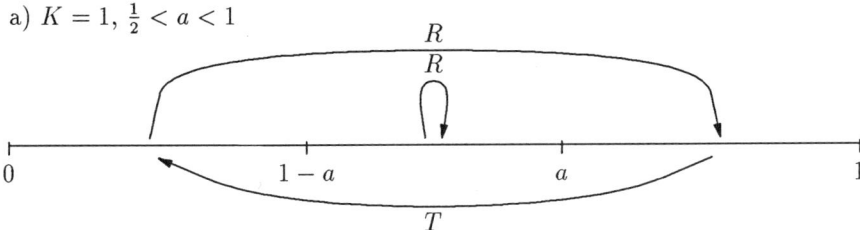

b) $K = 3$, $\frac{1}{4} < a < \frac{1}{3}$

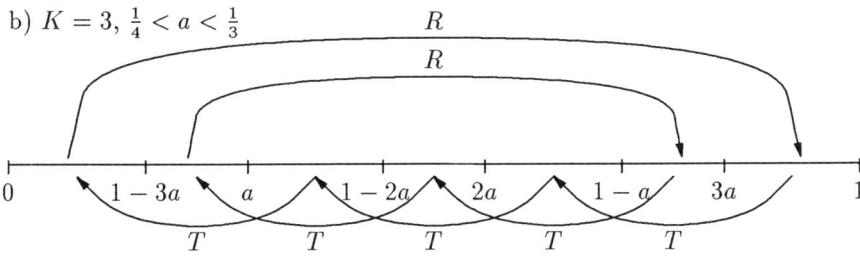

c) $\frac{1}{K+1} < a < \frac{1}{K}$

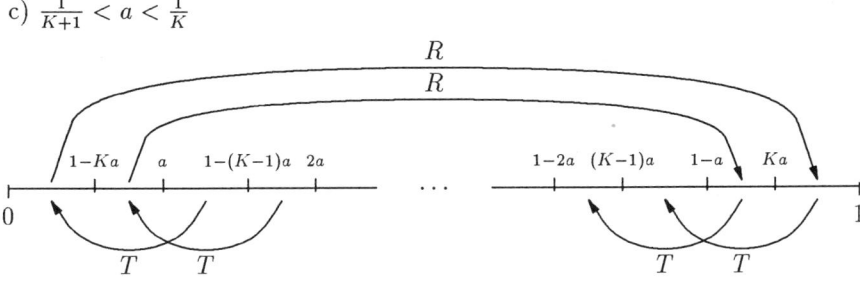

Figure 6.5: Partition examples for the analysis of Next Fit. Mappings shown are $R(x) = 1 - x$, $T(x) = x - a$.

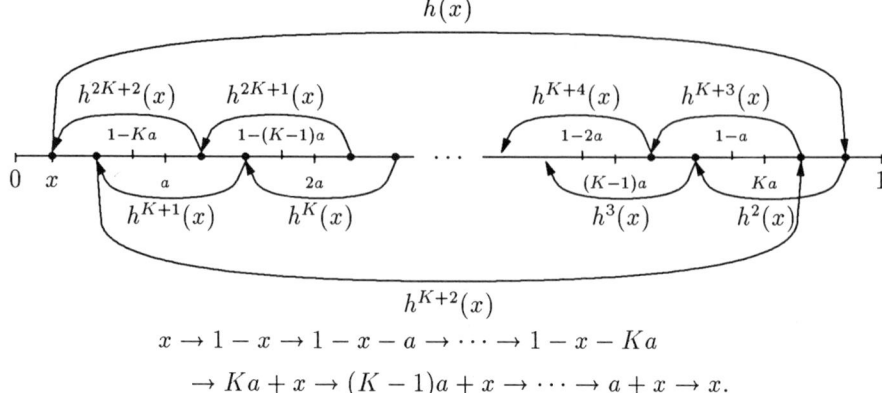

Figure 6.6: The mapping cycle.

back to the original interval $[0, 1 - Ka]$, but to a point $1 - Ka - x$, which is the reflection of x within $[0, 1 - Ka]$. Then the next $K + 1$ applications of $h(x)$, which are identical to the first $K+1$, map $1 - Ka - x$ back to the reflection of $1 - Ka - x$ in $[0, 1 - Ka]$, i.e., to the original point x.

We can now develop a linear system in the set of $2K + 2$ functions defined as the mappings $W(h^i(x))$, $0 \leq i \leq 2K + 1$, with x in a type-1 interval, and a linear system in the corresponding mappings $W(h^i(x))$, $0 \leq i \leq 2K - 1$, with x in a type-2 interval. Consider first the type-1 intervals and define

$$\phi_i(x) = W\left(h^{i-1}(x)\right) \quad \text{for } 1 \leq i \leq 2K + 2, \ 0 \leq x \leq 1 - Ka.$$

Then by (6.18) we have $\phi_1(x) = W(x)$ and

$$\phi_i(x) = \begin{cases} W\left(T^{i-2} R(x)\right) & \text{for } 2 \leq i \leq K + 2, \\ W\left(T^{i-K-3} R T^K R(x)\right) & \text{for } K + 3 \leq i \leq 2K + 2. \end{cases} \quad (6.19)$$

The symmetries illustrated in Figure 6.6 become

$$\phi_{i+K+1}(x) = \phi_i(1 - Ka - x) \quad \text{for } 1 \leq i \leq K + 1, \ 0 \leq x \leq 1 - Ka. \quad (6.20)$$

For example, with $K = 3$ we would have for $0 \leq x \leq 1 - 3a$,

$$\phi_1(x) = W(x), \qquad \phi_5(x) = W(1 - x - 3a),$$
$$\phi_2(x) = W(1 - x), \qquad \phi_6(x) = W(x + 3a),$$
$$\phi_3(x) = W(1 - x - a), \qquad \phi_7(x) = W(x + 2a),$$
$$\phi_4(x) = W(1 - x - 2a), \qquad \phi_8(x) = W(x + a).$$

6.3. ON-LINE LINEAR-TIME BIN PACKING

Noting that

$$\phi'_i(x) = \frac{d\phi_i(x)}{dx} = \begin{cases} -\dfrac{dW\left(h^{i-1}(x)\right)}{dx} & \text{for } i = 2, 3, \ldots, K+2, \\ \dfrac{dW\left(h^{i-1}(x)\right)}{dx} & \text{otherwise,} \end{cases}$$

we obtain the following linear system from (6.17):

$$\begin{aligned}
a\phi'_1(x) &= \phi_1(x) - \phi_2(x) + 1, & -a\phi'_{K+2}(x) &= \phi_{K+2}(x) - \phi_{K+3}(x) + 1, \\
-a\phi'_2(x) &= \phi_2(x) - \phi_3(x), & a\phi'_{K+3}(x) &= \phi_{K+3}(x) - \phi_{K+4}(x), \\
&\vdots & &\vdots \\
-a\phi'_K(x) &= \phi_K(x) - \phi_{K+1}(x), & a\phi'_{2K+1}(x) &= \phi_{2K+1}(x) - \phi_{2K+2}(x), \\
-a\phi'_{K+1}(x) &= \phi_{K+1}(x) - \phi_{K+2}(x), & a\phi'_{2K+2}(x) &= \phi_{2K+2}(x) - \phi_1(x).
\end{aligned} \qquad (6.21)$$

This system can be solved by standard techniques. The coefficient matrix **C** for (6.21) is the obvious generalization of the matrix shown below for $K = 3$.

$$\mathbf{C} = \frac{1}{a} \begin{bmatrix}
1 & -1 & 0 & 0 & 0 & 0 & 0 & 0 \\
0 & -1 & 1 & 0 & 0 & 0 & 0 & 0 \\
0 & 0 & -1 & 1 & 0 & 0 & 0 & 0 \\
0 & 0 & 0 & -1 & 1 & 0 & 0 & 0 \\
0 & 0 & 0 & 0 & -1 & 1 & 0 & 0 \\
0 & 0 & 0 & 0 & 0 & 1 & -1 & 0 \\
0 & 0 & 0 & 0 & 0 & 0 & 1 & -1 \\
-1 & 0 & 0 & 0 & 0 & 0 & 0 & 1
\end{bmatrix} \qquad (6.22)$$

Let $\Phi(x) = [\phi_1(x), \ldots, \phi_{2K+2}(x)]^T$, let **I** be the $(2K+2) \times (2K+2)$ unit matrix, and define γ as a column vector with a 1 in the first component, a -1 in the $(K+2)$th component, and 0's in the remaining $2K$ components. Then

$$\Phi'(x) = \mathbf{C}\Phi(x) + \gamma. \qquad (6.23)$$

By inspection of (6.22) for $K = 3$, the characteristic equation for the homogeneous system $\det[\mathbf{C} - \lambda\mathbf{I}] = 0$ generalizes to

$$[1 - (\lambda a)^2]^{K+1} = 1. \qquad (6.24)$$

The $2K+2$ roots are $\pm\sqrt{1 - r_j}/a$, where $r_1, \ldots, r_{K+1}$ are the $(K+1)$th roots of unity, with $r_{K+1} = 1$. Then we have a double root at 0 and single roots at $\pm\lambda_j$, where

$$\lambda_j = \frac{1}{a}\left(1 - \exp\left(\frac{j}{K+1}2\pi i\right)\right)^{1/2} \quad \text{for } 1 \leq j \leq K.$$

Then a fundamental set of solutions for the homogeneous system contains 1, x, and $e^{\pm\lambda_i x}$ ($1 \leq i \leq K$), where the 1 and x arise from the double root at 0. It is easily verified that adding x^2 to this set accounts for the nonhomogeneous term γ and gives us a fundamental set for (6.23), i.e., the ϕ_i are given by linear combinations of the $2K+3$ solutions in

$$\{1, x, x^2, e^{\lambda_1 x}, e^{-\lambda_1 x}, \ldots, e^{\lambda_K x}, e^{-\lambda_K x}\}. \tag{6.25}$$

An identical development leads to a linear system based on the type-2 intervals. Now we have an initial translation $x + 1 - Ka$, since the first type-2 interval starts at $1 - Ka$, so we define

$$\psi_i(x) = W\left(h^{i-1}(x+1-Ka)\right) \quad \text{for } 1 \leq i \leq 2K,\ 0 \leq x \leq (K+1)a - 1.$$

and hence $\psi_1(x) = W(x + 1 - Ka)$ and

$$\psi_i(x) = \begin{cases} T^{i-2} R(x+1-Ka) & \text{for } 2 \leq i \leq K+1, \\ T^{i-K-2} R T^{K-1} R(x+1-Ka) & \text{for } K+2 \leq i \leq 2K, \end{cases} \tag{6.26}$$

with the symmetries

$$\psi_{i+K}(x) = \psi_i((K+1)a - 1 - x) \quad \text{for } 1 \leq i \leq K. \tag{6.27}$$

By repeating the earlier analysis, the functions $\psi_i(x)$, $1 \leq i \leq 2K$, are obtained as linear combinations of the functions in the fundamental set

$$\{1, x, x^2, e^{\mu_1 x}, e^{-\mu_1 x}, \ldots, e^{\mu_{K-1} x}, e^{-\mu_{K-1} x}\}, \tag{6.28}$$

where $\mu_j = \sqrt{1 - s_j/a}$ and $s_1, s_2, \ldots, s_{K-1}$ along with $s_K = 1$ are the Kth roots of unity.

As the last step in determining the ϕ_i and ψ_i, the unknown coefficients in the linear combinations of the functions in the respective sets (6.25) and (6.28) have to be evaluated. In general, these are calculated as the solutions to linear systems determined by the boundary conditions:

1. $W(0) = 0$, $W(1) = 1$.

2. $W(x)$ is continuous at the boundaries between the type-1 and type-2 intervals.

6.3. ON-LINE LINEAR-TIME BIN PACKING

As illustrated later, these are converted to conditions on the ϕ_i and ψ_i by (6.19), (6.20), (6.26), and (6.27). Explicit forms for the coefficients as functions of K and a do not appear to be known, so we will not pursue the general case further, except to note that, once the ϕ_i and ψ_i are known, $W(x)$ can be assembled from $\phi_1(x)$, $\psi_1(x)$, and the first $K+1$ and K functions, respectively, in (6.19) and (6.26).

Explicit solutions can be expected to have rather complicated forms. This is true even for the case $K = 1$, $1/2 < a < 1$, which we now treat in detail. As shown in Figure 6.5, the subintervals are $[0, 1-a]$, $[1-a, a]$, and $[a, 1]$; as noted earlier the single type-2 interval $[1-a, a]$ maps into itself under $R(x) = 1-x$. From (6.19) and (6.26) we have

$$\left.\begin{aligned}\phi_1(x) &= W(x) \\ \phi_2(x) &= W(1-x) \\ \phi_3(x) &= W(1-x-a) \\ \phi_4(x) &= W(x+a)\end{aligned}\right\} \quad \text{for } 0 \leq x \leq 1-a,$$

$$\left.\begin{aligned}\psi_1(x) &= W(1-a+x) \\ \psi_2(x) &= W(a-x)\end{aligned}\right\} \quad \text{for } 0 \leq x \leq 2a-1,$$

so from (6.25) and (6.28)

$$\phi_i(x) = \alpha_{i1} + \alpha_{i2}x + \alpha_{i3}x^2 + \alpha_{i4}e^{\lambda x} + \alpha_{i5}e^{-\lambda x} \quad \text{for } 1 \leq i \leq 4,$$

with $\lambda = \sqrt{2}/a$, and

$$\psi_i(x) = \beta_{i1} + \beta_{i2}x + \beta_{i3}x^2.$$

The boundary conditions give

$W(0) = 0:\qquad \phi_1(0) = \phi_3(1-a) = 0,$

$W(1) = 0:\qquad \phi_2(0) = \phi_4(1-a) = 1,$

$W\big((1-a)^-\big) = W\big((1-a)^+\big):\qquad \phi_1(1-a) = \psi_1(0) = \psi_2(2a-1) = \phi_3(0),$

$W(a^-) = W(a^+):\qquad \psi_1(2a-1) = \phi_4(0) = \phi_2(1-a) = \psi_2(0).$

A straightforward but tedious elimination, and then substitution into

$$W(x) = \begin{cases} \phi_1(x) & \text{for } 0 \leq x \leq 1-a, \\ \psi_1(x-1+a) & \text{for } 1-a \leq x \leq a, \\ \phi_2(1-x) & \text{for } a \leq x \leq 1, \end{cases}$$

then yields the solution

$$W(x) = \begin{cases} u(x) + \dfrac{w}{2}\left[\tanh w + \dfrac{\sinh(\lambda x - w)}{\cosh w}\right] & \text{for } 0 \le x \le 1-a, \\ 2u(x) + w\tanh w & \text{for } 1-a \le x \le a, \\ u(x) + \dfrac{1}{2} + \dfrac{w}{2}\left[\tanh w - \dfrac{\cosh(\lambda x - \beta)}{\cosh w}\right] & \text{for } a \le x \le 1, \end{cases} \quad (6.29)$$

where

$$u(x) = \frac{x^2 - (1-a)x}{2a^2}, \quad w = \frac{1-a}{a\sqrt{2}}, \quad \beta = \lambda - w - \ln(1+\sqrt{2}).$$

Substitution into (6.15) and (6.14) then proves

Theorem 6.10 ([Karm82]) *For the distribution $U(0,a)$, $1/2 < a \le 1$, of item sizes, we have*

$$\mathsf{E}[\mathrm{NF}(L_n)] \sim \rho n \quad \text{as } n \to \infty,$$

where

$$\rho = \frac{1}{12a^3}\left(15a^3 - 9a^2 + 3a - 1\right) + \sqrt{2}\left(\frac{1-a}{2a}\right)^2 \tanh\left(\frac{1-a}{\sqrt{2a}}\right). \quad (6.30)$$

By Theorem 5.5, the interval $[0,a]$, $0 < a \le 1$, allows perfect packing, so we can also write

$$\frac{\mathsf{E}[\mathrm{NF}(L_n)]}{\mathsf{E}[\mathrm{OPT}(L_n)]} \sim \frac{\rho n}{na/2} = 2\rho/a \quad \text{as } n \to \infty.$$

The reciprocal, $\eta = a/2\rho$, measures the asymptotic efficiency of NF packings. Intuitively, one might expect that the smaller the value of a, the higher the efficiency. Thus it is interesting to learn that, in fact, η does not decrease monotonically as a increases [Karm82]. A numerical calculation using (6.30) shows that η attains a minimum at $a \approx 0.841$ in the region $1/2 < a \le 1$. See Figure 6.7.

6.3.2 Deviation from the expected behavior

In [RT87] it is shown that the monotonicity property (Section 2.6) and a martingale inequality can be used to bound the tail of the distribution of

6.3. ON-LINE LINEAR-TIME BIN PACKING

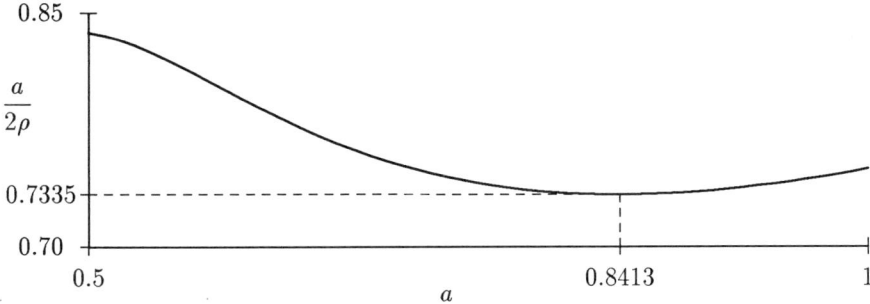

Figure 6.7: The asymptotic efficiency of NF packings.

the variation from the mean $|H(L_n) - \mathsf{E}[H(L_n)]|$. We need one additional term: say a heuristic is *k-conservative* if adding an item anywhere in the list of items cannot increase the number of bins used by more than k. Note that this property is not implied by monotonicity; in particular, the algorithm which produces an optimum packing if there are k' or fewer items, but packs them one per bin if there are more than k' items, is monotonic but not k-conservative for any $k < k'$. It is, however, easy to see that OPT is 1-conservative and Next Fit is 2-conservative.

Now suppose that H is monotonic and 1-conservative and let $L_n^{(i)}$ denote list L_n with X_i removed. Note that then

$$H(L_n^{(i)}) \leq H(L_n) \leq H(L_n^{(i)}) + 1 \quad \text{for all } L_n. \tag{6.31}$$

Define the random variables

$$D_i = \mathsf{E}[H(L_n) \mid L_i] - \mathsf{E}[H(L_n) \mid L_{i-1}] \quad \text{for } 1 \leq i \leq n.$$

(Here the notation $\mathsf{E}[H(L_n) \mid L_i]$ means the random variable found by averaging over all but the first i elements of L_n; thus, in particular, $\mathsf{E}[H(L_n) \mid L_n]$ is just the random variable $H(L_n)$, and $\mathsf{E}[H(L_n) \mid L_0] = \mathsf{E}[H(L_n)]$.) The sum of the D_i telescopes and yields $\sum_{i=1}^n D_i = H(L_n) - \mathsf{E}[H(L_n)]$. Clearly, $\mathsf{E}[D_i] = 0$, $1 \leq i \leq n$, so the sequence $D_1, \ldots, D_n$ constitutes a *martingale difference sequence*. Moreover, from (6.31), it is easy to verify that $|D_i| \leq 1$, $1 \leq i \leq n$. We can now apply Azuma's inequality for martingale difference sequences. (This result can be found in [Stou74, Lemma 4-2-3 and Exercise 4-2-2]; alternatively, one can apply Theorem 2.5 (page 19).) By this result we have

$$\Pr\{|H(L_n) - \mathsf{E}[H(L_n)]| > t\} = \Pr\left\{\left|\sum_{i=1}^n D_i\right| > t\right\} \leq 2e^{-t^2/2n}.$$

Thus the probability that the variation from the mean exceeds a multiple of $\sqrt{n}$ can be made as small as desired.

A heuristic H enjoys the *subadditive* property if

$$H(L_m \circ L_n) \leq H(L_m) + H(L_n), \qquad (6.32)$$

where $L_m \circ L_n$ denotes the concatenation of L_m and L_n. It is easily seen that both NF and OPT have the subadditive property. From the theory of subadditive processes [King76], we have $\lim_{n\to\infty} H(L_n)/n$ exists almost surely; let c be this limit. Thus, since $H(l_n)/n$ always lies in $[0,1]$, we also have have $\lim_{n\to\infty} \mathsf{E}[H(L_n)]/n = c$, so that for any $\epsilon > 0$, $|\mathsf{E}[H(L_n)]/n - c|$ will become an arbitrarily small fraction of ϵ for large enough n. Hence we can use (6.32) to obtain a stronger asymptotic result:

Theorem 6.11 [RT87] *For a 1-conservative subadditive monotonic heuristic H, there exists a constant $c > 0$ such that*

$$\forall \alpha > 2 \ \forall \epsilon > 0 \ \exists n_0 \text{ such that } \forall n \geq n_0, \Pr\left\{ \left| \frac{H(L_n)}{n} - c \right| > \epsilon \right\} \leq 2e^{-n\epsilon^2/\alpha}.$$

Note that this result can easily be extended to algorithms that are k-conservative by a simple scaling. For example, $\text{NF}(L_n)$ is not 1-conservative, but $\frac{1}{2}\text{NF}(L_n)$ is, so as noted in [RT87] as a corollary we can obtain a similar bound for Next Fit.

6.3.3 The HARMONIC algorithm

At a sacrifice in the simplicity of NF, the HARMONIC algorithm improves on the expected wasted space while retaining the linear-time property. HARMONIC partitions the items of L_n on-line into subintervals $(1/2, 1]$, $(1/3, 1/2], \ldots, (1/M, 1/(M-1)], (0, 1/M]$ with $M \geq 1$ a given parameter. For $m < M$, *type-m items* are those with sizes in $(1/(m+1), 1/m]$; items with sizes in $(0, 1/M]$ are *type-M*. For each m, $1 \leq m \leq M$, HARMONIC forms a separate NF packing of the type-m items; call the bins used in this packing *type-m bins*. In particular, since we are using NF, for each m there is only one active type-m bin at any given time. Formally, we initialize the set of active bins to be empty, and then iterate the following process for each X in L_n. Let m be the type of X. If there is no active type-m bin (which occurs if X is the first type-m item), then we place X into a new active type-m bin. If there is an active type-m bin, and X fits into it, we pack X into this bin. If there is an active type-m bin, but X does not fit into it, we close this bin,

6.3. ON-LINE LINEAR-TIME BIN PACKING

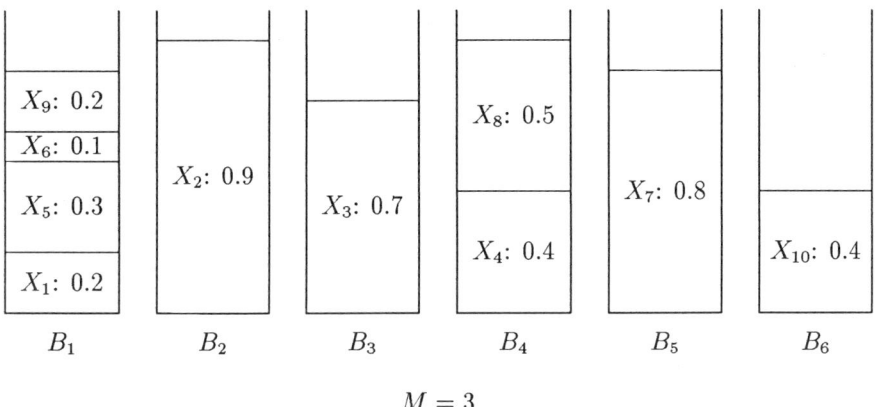

Figure 6.8: An example for the HARMONIC algorithm. Here $L_{10} = (X_1, X_2, \ldots, X_{10}) = (0.2, 0.9, 0.7, 0.4, 0.3, 0.1, 0.8, 0.5, 0.2, 0.4)$, and B_5, B_6, and B_1 are active type-1, -2, and -3 bins, respectively.

and pack X into a new active type-m bin. Figure 6.8 shows an example. Note that there is no upper bound to the number of items in a type-M bin, but when such such a bin is full, it has at least M items; full type-m bins for $1 \le m \le M-1$ have exactly m items. Note also that HARMONIC reduces to NF when $M = 1$. Below we assume $M \ge 2$.

The average case for HARMONIC is easily estimated for general item-size distributions (see [LL87]). Let $p_m = \Pr\{\frac{1}{m+1} < X \le \frac{1}{m}\}$ for $1 \le m \le M-1$, and $p_M = \Pr\{0 < X \le 1/M\}$, so that $\sum_{1 \le m \le M} p_m = 1$ and np_m is the expected number of type-m items for $1 \le m \le M$. Then the expected number of type-m bins, $1 \le m \le M$, in a HARMONIC packing is at most $np_m/m + 1$, and we get

$$\mathsf{E}[\mathrm{HARMONIC}(L_n, M)] \le n \left(\sum_{m=1}^{M} \frac{p_m}{m} \right) + M.$$

Let $R_M = \lim_{n \to \infty} \mathsf{E}[\mathrm{HARMONIC}(L_n, M)]/\mathsf{E}[\mathrm{OPT}(L_n)]$. Then, since $n\mathsf{E}[X]$ is a lower bound on $\mathsf{E}[\mathrm{OPT}(L_n)]$, we have

$$R_M \le \frac{1}{\mathsf{E}[X]} \sum_{m=1}^{M} \frac{p_m}{m},$$

with equality for perfectly packable distributions; indeed, for such distributions,

$$R_M \sim \frac{1}{\mathsf{E}[X]} \sum_{m=1}^{\infty} \frac{p_m}{m} \quad \text{as} \quad M \to \infty. \tag{6.33}$$

For the item-size distribution $U(0,1)$, we have $p_m = \frac{1}{m} - \frac{1}{m+1}$ for $1 \leq m < M$, and $p_M = 1/M$. Substitution into (6.33) gives $R_M \sim \pi^2/3 - 2 = 1.289868\ldots$ as $M \to \infty$, which may be compared with $4/3 = 1.333\ldots$ for NF.

For the distribution $U(0,1)$, the expected wasted space in bins with items larger than $1/2$ is $\frac{n}{2} \cdot \frac{1}{4} = \frac{n}{8}$, which is a large majority of the expected wasted space (about $0.1449n$). Thus attempts to pack even one more item along with items larger than $1/2$ can lead to marked improvements in packing efficiency. Ramanan and Tsuga [RT89a] have analyzed such extensions to HARMONIC.

6.3.4 On-line matching

The fraction of the space wasted by NF and HARMONIC remains bounded away from zero even in the asymptotic limits. Hoffmann [Hoff82] developed an on-line, linear-time version of MATCH (see Section 5.1) that is asymptotically optimal in the sense that the ratio of wasted space to occupied space tends to zero as n and M, a parameter of the algorithm, tend to infinity in the order $M \ll n$. Lee and Lee [LL87] subsequently devised a variant that improved the worst-case behavior while retaining the asymptotic optimality. We present a somewhat simpler variant, called ONLINEMATCH; the analysis is similar and leads to a similar result. (The algorithm and result presented here are closely related to the Interval First Fit of [CG86], which achieves expected wasted space $O(n^{2/3})$.) In particular, the key to the analysis is a fundamental property of symmetric random walks.

For a given even parameter $M \geq 4$, ONLINEMATCH classifies items into M types: an item in $(\frac{m-1}{M}, \frac{m}{M}]$ is of type m, $1 \leq m \leq M$. Two items, one of type m and one of type $M - m$, $1 \leq m < M$, are called *companions*; they can fit together (be matched) in a single bin. Note that companions are not defined for type-M items and that any two type-$M/2$ items are companions. As described below, the packings of ONLINEMATCH can be divided, in general, into three sets of bins, one set containing only type-M items, another set comprising an NF packing of items less than $1/2$, and a third set comprising a partial MATCH packing.

Depending on its type and the current packing, the next item, say, X, of L_n is packed according to the following case analysis. Let X be of type m.

6.3. ON-LINE LINEAR-TIME BIN PACKING

1. $M/2 < m \leq M$: pack X into the next (lowest-indexed) empty bin.

2. $m = M/2$: if there is a bin B with a type-$M/2$ item alone in it, then pack X with its companion in B; otherwise, pack X into the next empty bin.

3. $1 \leq m < M/2$: if there is a bin B with a companion of X alone in it, then pack X into B. If no such bin exists, then pack X into the current NF bin, if such a bin exists and X fits in it; otherwise, pack X into the next empty bin (which then becomes the current NF bin).

According to this algorithm, an item $X < 1/2$ of type $m < M/2$ is matched with another item only if there is a bin with a single item $X' > 1/2$ of type $M - m$ already packed and waiting to be matched with a companion. Otherwise, such items are packed according to Next Fit. It is easy to verify the worst-case bound

$$\text{ONLINEMATCH}(L_n, M) \leq 2\,\text{OPT}(L_n).$$

We now show that ONLINEMATCH is asymptotically optimal in terms of the expected wasted space.

Let n_m be the number of type-m items in L_n, so that $\mathsf{E}[n_m] = n/M$, $1 \leq m \leq M$. In the final packing, the expected number of bins containing items of type m is n/M for each $M/2 < m \leq M$, and at most $n/2M + 1$ for $m = M/2$. Let K_n be the total number of unmatched items (packed by NF) of types $1 \leq m < M/2$. Since these items can be packed at least two per bin, they require at most $\frac{1}{2}\mathsf{E}[K_n] + 1$ bins on the average, and we have

$$\mathsf{E}[\text{ONLINEMATCH}(L_n, M)] \leq \frac{M}{2}\frac{n}{M} + \frac{n}{2M} + 1 + \frac{1}{2}\mathsf{E}[K_n] + 1$$

$$\leq \frac{n}{2}\left(1 + \frac{1}{M}\right) + \frac{1}{2}\mathsf{E}[K_n] + 2. \qquad (6.34)$$

Let $K_{m,n}$ be the number of unmatched items of type m, $1 \leq m < M/2$, so that $K_n = \sum_{m=1}^{M/2-1} K_{m,n}$. Let m be any integer in the range $1 \leq m < M$. The analysis of $K_{m,n}$ is similar to the methods of proof used in Theorem 5.1 and Theorem 5.5, so we only briefly outline it here. The distribution of the set of items of type-m and type-$(M-m)$ can be described as follows: we first pick an integer k according to a binomial distribution giving the number of successes in n trials each with success probability $2/M$, and then we pick k i.i.d. items uniformly from the union of intervals

$$\left(\frac{m-1}{M}, \frac{m}{M}\right) \cup \left(\frac{M-m-1}{M}, \frac{M-m}{M}\right).$$

If we concentrate our attention on the manner in which these items are packed, we see that, by an analysis somewhat like that in Section 5.1, for a given k, the expected number of unmatched items of type-m is distributed as the maximum excess of heads over tails in k flips of a fair coin; thus the expectation is bounded by $c\sqrt{k}$ for some absolute constant c. However, k itself is a random variable with mean $2n/M$, so by an application of Jensen's inequality like that in Theorem 5.5, we may conclude that the number of unmatched type-m items is

$$\mathsf{E}[K_{m,n}] \leq c\sqrt{2n/M}.$$

Hence

$$\mathsf{E}[K_n] \leq \sum_{m=1}^{M/2-1} K_{m,n} \leq \left(\frac{M}{2} - 1\right) c\sqrt{2n/M} \leq c\sqrt{nM/2}. \qquad (6.35)$$

Finally, by (6.34) the ratio of expected wasted space to expected occupied space is bounded by

$$\frac{n/2M + \frac{1}{2}c\sqrt{nM/2} + 2}{n/2},$$

so we obtain

Theorem 6.12 *For the item-size distribution $U(0,1)$,*

$$\frac{\mathsf{E}[\mathrm{ONLINEMATCH}(L_n, M)] - \mathsf{E}[\mathrm{OPT}(L_n)]}{\mathsf{E}[\mathrm{OPT}(L_n)]} \leq \frac{1}{M} + C\sqrt{M/n}$$

for some absolute constant C. In particular,

$$\lim_{M \to \infty} \lim_{n \to \infty} \frac{\mathsf{E}[\mathrm{ONLINEMATCH}(L_n, M)]}{\mathsf{E}[\mathrm{OPT}(L_n)]} = 1.$$

6.3.5 On-line packing with limited active bins

Recall that we have defined a bin to be active if it has already received at least one item, but is still available as a candidate to receive additional items; in particular, NF satisfies the condition that it never has more than one active bin (Section 1.4.2). It is interesting to see how far we can push the performance of on-line packing algorithms when the number of active bins must be bounded by a given constant throughout the running of the algorithm.

6.3. ON-LINE LINEAR-TIME BIN PACKING

If we restrict the number of active bins to be at most 1, it might at first glance seem that NF is the best we can do. However, a variation called *Smart Next Fit* (SNF), introduced and analyzed in [Rama89], can bring about a substantial improvement in the average-case behavior for some distributions. When a new item x arrives to be packed, we pack it into the current bin B if it fits, as in NF. If it does not fit, we pack it into a new bin B'. We then let whichever of B and B' has more remaining capacity become the new current bin, and close the other of B and B'. Note that since we can decide which of B and B' will become the new current bin before we actually pack x into B', this algorithm can be implemented so that it only has one active bin according to the above definition. (Note that if all item sizes are bounded by 1/2, SNF is equivalent to ordinary NF.)

The analysis of this algorithm provides another example of a situation in which the techniques of Karmarkar [Karm82] can be applied. Figure 6.9 illustrates the argument; there the possible outcomes are plotted as regions in a space whose coordinates are the possible values of the level z of the current bin, and the size x of the next item to be packed. The situations that lead to a new current-bin level of u are shown as a dotted line. Let the distribution of the items be $F(x)$, as usual, with density $f(x)$. Assuming we are at steady state, let the cumulative distribution of the size z of the current bin be $\Phi(z)$, with density $\phi(z)$. Then one sees from the figure that the density after this next item is packed satisfies

$$\phi(u) = \int_0^u \phi(z) f(u-z)\, dz + f(u) \int_{\max(u,1-u)}^1 \phi(z)\, dz$$

$$+ \phi(u) \int_{\max(u,1-u)}^1 f(x)\, dx$$

$$= \int_0^u \phi(z) f(u-z)\, dz + f(u)\bigl[1 - \Phi(\{\max(u, 1-u)\})\bigr]$$

$$+ \phi(u)\bigl[1 - F(\max\{u, 1-u\})\bigr].$$

Assuming a distribution uniform over $(0, b)$, $b \leq 1$, this integral equation can readily be converted to four simultaneous differential equations, corresponding to values of Φ over the intervals $[0, 1-b]$, $[1-b, 1/2]$, $[1/2, b]$, $[b, 1]$; we omit the details of the solution, but display some results. The packing efficiency of SNF is compared to that of NF in Figure 6.10; as can be seen, the two algorithms have identical efficiencies at $b = 1/2$ (as expected), but SNF is substantially better at $b = 1$.

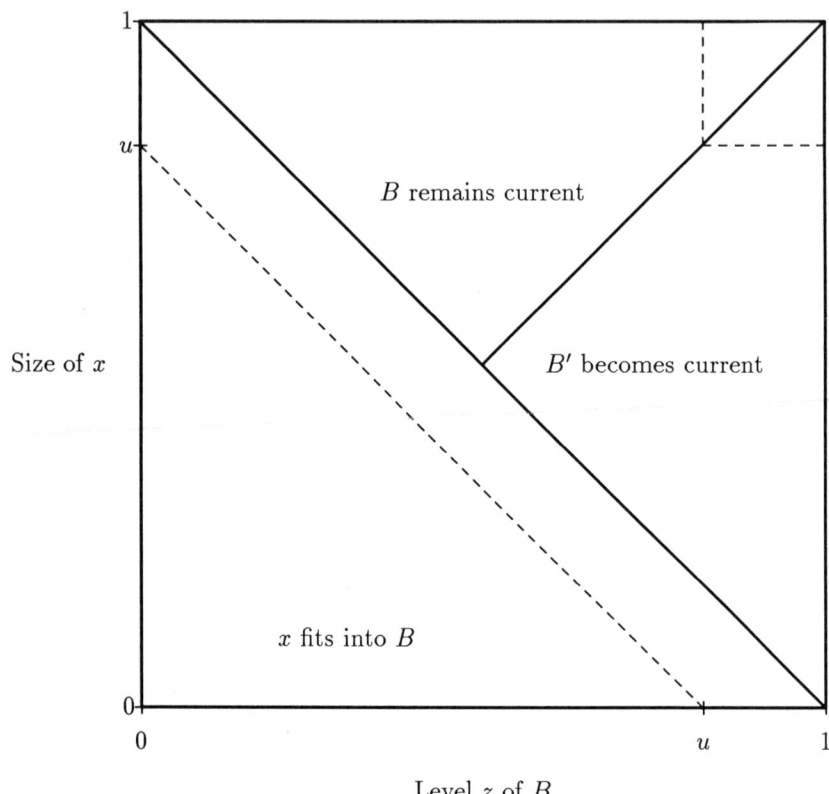

Figure 6.9: Illustration of the argument leading to an integral equation for Smart Next Fit. Assume that B is the current bin, and we are about to pack an item x. Below the main diagonal, x fits into B. Above the main diagonal, we need to allocate a new bin B'. The cases in which the level of the new current bin is equal to u are shown as dashed lines.

6.3. ON-LINE LINEAR-TIME BIN PACKING

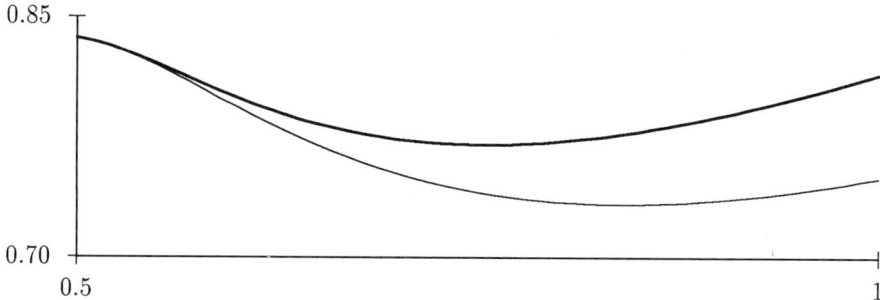

Figure 6.10: Comparison of the efficiency of Next Fit and Smart Next Fit. Assuming the item sizes are i.i.d. uniform over $[0, b]$, the asymptotic average level of the packed bins is plotted versus b. The bold curve is for SNF; the lighter curve is for NF, for comparison.

One might wonder whether we can achieve asymptotic expected optimality with an algorithm that requires only a bounded number of active bins. The answer is no, as the theorem below will show.

Theorem 6.13 ([CSa]) *Let H be a bin-packing heuristic that is limited to r active bins. Then for item sizes i.i.d. over $[0, 1]$, the expected wasted space in a packing of n items is*

$$E[H(L_n)] - \frac{n}{2} \geq \frac{n}{16(r+1)}.$$

Proof. Say that a bin B is w-tight at some point in time if the unused capacity in B is at most w; here w is a parameter whose value will be chosen later. To facilitate the counting of such bins, we will define an item X_i to be w-tight if, just after X_i is packed, the bin into which it is packed is w-tight.

Suppose we are about to pack X_i. Let $\mathcal{A}_i$, with $|\mathcal{A}_i| = s \leq r + 1$, be the current set of active bins together with the next available empty bin. Thus we are assured that X_i will be packed into some element of $\mathcal{A}_i$. Let l_j, $1 \leq j \leq s$, be the current levels of the bins in $\mathcal{A}_i$; in particular, assuming l_s corresponds to the empty bin, we have $l_s = 0$. Now, by definition, a necessary condition for X_i to be w-tight is that there be some bin B in $\mathcal{A}_i$ such that X_i fills B to within w of its capacity. Hence, using Boole's inequality, we have

$$\Pr\{X_i \text{ is } w\text{-tight}\} \leq \Pr\{\exists j,\ 1 \leq j \leq s,\ 1 - w \leq l_j + X_i \leq 1\}$$
$$\leq \sum_{j=1}^{s} \Pr\{1 - w \leq l_j + X_i \leq 1\}$$

$$= \sum_{j=1}^{s} \Pr\{X_i \in [1 - w - l_j, 1 - l_j]\}$$
$$\leq \sum_{j=1}^{s} w = ws \leq w(r+1).$$

Summing over all items, we can bound the expected number of w-tight items by $nw(r + 1)$; note that this is also a bound on the expected number of w-tight bins at the end of the packing, since the last item to go into a w-tight bin must be a w-tight item. The total expected number of bins is at least $n/2$, since this is the expected total size of the items. Thus, on average, at least $n/2 - nw(r+1)$ bins waste at least w of their capacity, so the expected total waste is at least $w\bigl(n/2 - nw(r+1)\bigr)$. It is routine to see that this is maximized at $w = \bigl(4(r+1)\bigr)^{-1}$, giving the bound of the theorem. ∎

Actually, we do not need to restrict the number of active bins in order to achieve such a bound—the crucial condition needed is that whenever we are about to pack an item X_i, there is always some set $\mathcal{A}_i$ of bins, that does not depend on X_i, such that $|\mathcal{A}_i| \leq r + 1$, and we are guaranteed that X_i will be packed into an element of $\mathcal{A}_i$. Note that $\mathcal{A}_i$ can be allowed to depend on $X_1, X_2, \ldots, X_{i-1}$. The statement of the theorem in [CSa] uses this more general context.

Chapter 7

Packings in Two Dimensions

In two dimensions we study *strip packing* and its variant, *two-dimensional bin packing*. (For a recent survey on this subject see [CS90].) In the first problem, the rectangles in a given list L_n have sides no greater than 1; they must be packed into a semi-infinite strip of width 1. Packings of the strip, which we draw vertically by convention, must be such that (i) rectangles do not overlap each other or the boundaries of the strip; (ii) the rectangles are packed with their sides parallel to the sides of the strip; and (iii) the height of the packing (the maximum height of the top of any rectangle) is minimized. Figure 7.1 shows an example. The dimensions of a rectangle R_i will be denoted by X_i and Y_i, for the horizontal and vertical sides, respectively. Unless specified otherwise, rectangles must be packed in their given orientations, i.e., no rotations are allowed.

Now suppose horizontal boundaries are also placed at the integer heights of the strip, and no rectangle is allowed to overlap any of these boundaries. Then we have *two-dimensional bin packing*; the square region between successive integer heights is a "bin." In this problem, the "size" of the packing is taken to be the ceiling of the height of the packing, i.e., the number of bins used.

We present the analysis of bounds on optimum packings and the performance of several heuristics. We begin with off-line algorithms and then study on-line algorithms. The terms "item" and "rectangle" will be used interchangeably in this chapter.

7.1 Off-line algorithms

Our first problem is strip packing when the items are squares. Tight bounds are derived for this special case, and serve as an example of the "bound that

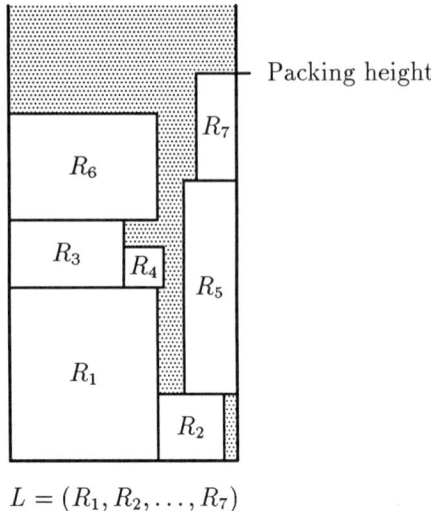

Figure 7.1: A strip-packing example.

usually holds" technique of Section 2.5. Next, we study bounds for optimum strip packings of rectangles. Finally, we present and analyze a matching algorithm for two-dimensional bin packing.

7.1.1 Packing squares into a strip

In this section, we study the problem of packing squares into a strip; this problem is interesting partly because of the biased random walk that arises in the analysis. A square (X_i, X_i) is denoted simply by X_i. A phrase such as "a square X_i in $[0,1]$" has the obvious meaning of a square all of whose sides are in $[0,1]$. Let the X_i be independently and uniformly distributed over $[0,1]$. Figure 7.2 shows an example. A brief study of the sample L_n will convince the reader that the packing in Figure 7.2 is optimum.

Let the sum of the square sizes exceeding $1/2$ be denoted by

$$H_{1/2} = H_{1/2}(L_n) = \sum_{i: X_i > 1/2} X_i. \tag{7.1}$$

The average total area of the squares is

$$n\mathsf{E}[X_i^2] = n \int_0^1 x^2 dx = \frac{n}{3},$$

7.1. OFF-LINE ALGORITHMS

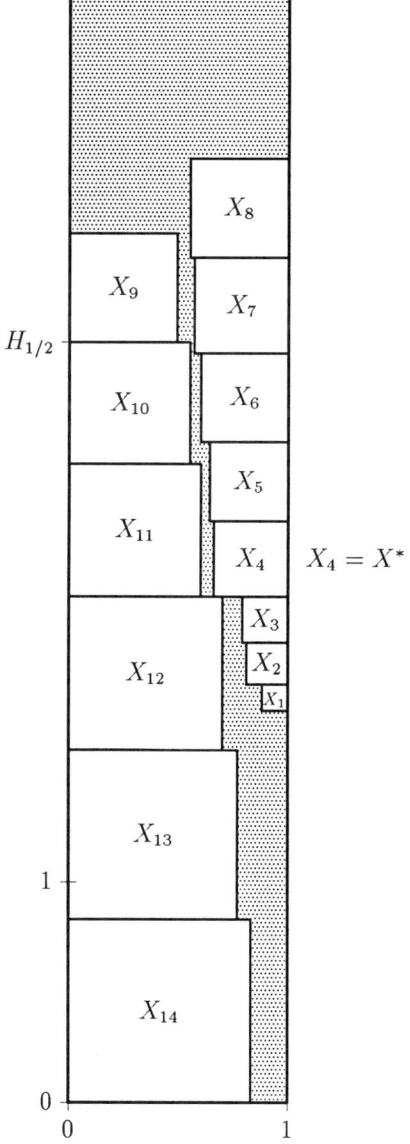

$L_n = (X_1, X_2, \ldots, X_{14})$
$= (0.12, 0.19, 0.21, 0.34, 0.36, 0.40, 0.43, 0.45, 0.49, 0.55, 0.60, 0.70, 0.77, 0.83)$

Figure 7.2: A square-packing example.

but

$$\mathsf{E}[H_{1/2}] = \frac{n}{2} \cdot \frac{3}{4} = \frac{3n}{8}. \tag{7.2}$$

Thus, since $\mathrm{OPT}(L_n) \geq H_{1/2}$, even an optimum algorithm must waste at least $3n/8 - n/3 = n/24$ space on the average. In this sense, packings of squares drawn uniformly from $[0,1]$ are not particularly efficient. A more detailed explanation of this fact is as follows. Opposite each square X in $[1-a, 1]$ we could pack $\lfloor (1-a)/a \rfloor$ squares no larger than a without extending beyond the bottom or top of X. However, in our probabilistic model the expected numbers of squares in $[0, a]$ and in $[1-a, 1]$ are equal. Then for a moderately smaller than $1/2$ we can expect to have more space than we need for the efficient packing of squares in $[0, a]$, even when we allow for probable variations in the square sizes of L_n. Indeed, we prove later that if all squares in L_n with sizes in $[0, 1/3]$ are removed, then with a probability that quickly approaches 1 as $n \to \infty$, the optimum packing height remains unchanged. Note that the squares no larger than $1/3$ in Figure 7.2 (namely, X_1, X_2, X_3) could be removed without affecting the height of the packing.

We now look for a lower bound better than $\mathrm{OPT}(L_n) \geq H_{1/2}$ that concentrates on the larger squares, in hopes that we can prove that the bound "usually holds."

Consideration of Figure 7.2, and the fact that no two squares with sizes above $1/2$ can be alongside each other, suggests the following algorithm for packing squares.

Algorithm A

1. *Stack the squares with widths greater than $1/2$ along the left edge of the strip in order of decreasing width.*

2. *Starting at height $H_{1/2}$, stack the remaining squares along the right edge of the strip in order of increasing width.*

3. *Slide the stack on the right edge down until it rests on the bottom of the strip, or a square in the right stack comes in contact with a square in the left stack, whichever occurs first.*

4. *Repack the squares lying entirely above $H_{1/2}$ into two stacks, one against the left edge of the strip and the other against the right edge. Pack these squares in decreasing order of size, with the ith square being placed on the shorter of the two stacks created by the first $i-1$ of these squares.*

7.1. OFF-LINE ALGORITHMS

Figure 7.3 shows the packing of L_n in Figure 7.2 after the first three steps of Algorithm A. After Step 4 we obtain the packing in Figure 7.2. Note that Step 4 is an adaptation of the two-processor LPT rule defined in Section 1.4.1.

To formalize ideas, it is convenient to define $\delta(y)$ as the total height of squares in $[1/2 - y, 1/2]$ minus the total height of squares in $(1/2, 1/2 + y]$, i.e., $\delta(0) = 0$ (almost surely) and

$$\delta(y) = \sum_{i: 1/2-y \leq X_i \leq 1/2} X_i - \sum_{i: 1/2 < X_i \leq 1/2+y} X_i \quad \text{for } 0 < y \leq 1/2. \tag{7.3}$$

Let

$$\Delta = \Delta(L_n) = \max_{0 \leq y \leq 1/2} \delta(y), \tag{7.4}$$

and define X^* as the largest value such that the maximum in (7.4) is achieved at $y = 1/2 - X^*$. Δ is illustrated in Figure 7.3, where $X^* = X_4$. Note that Δ can never be negative, since $\delta(0) \geq 0$. It is quite possible, however, that Δ is equal to 0, in which case X^* must be $1/2$; for example, consider the list in Figure 7.3 but with squares X_6, X_7, X_8, and X_9 removed.

The following lemma gives a lower bound on $\text{OPT}(L_n)$ in terms of Δ and X^*; as we will soon see, $\text{OPT}(L_n)$ is usually close to this bound.

Lemma 7.1 *For any $y \in [0, 1/6)$, we have $\text{OPT}(L_n) \geq H_{1/2} + \delta(y)/2$. This means that, if $X^* > 1/3$, then*

$$\text{OPT}(L_n) \geq H_{1/2} + \Delta/2.$$

Proof. Fix $y \in [0, 1/6)$ and let S_1, S_2, and S_3 be the total sizes of items in $[1/2 - y, 1/2]$, $(1/2, 1/2 + y)$, and $(1/2 + y, 1]$, respectively. Note that $H_{1/2} = S_2 + S_3$ and $\delta(y) = S_1 - S_2$. Now if we consider only squares in $[1/2 - y, 1]$, it is clear that no square fits alongside those in S_3, and that those in $S_1 \cup S_2$ can fit at most two abreast. Hence

$$\text{OPT}(L_n) \geq \frac{S_1 + S_2}{2} + S_3 = (S_2 + S_3) + \frac{S_1 - S_2}{2} = H_{1/2} + \delta(y)/2. \quad \blacksquare$$

The main theorem of this section will show that $X^* > 1/3$ holds, and hence the lower bound holds, with very high probability. We now show that Algorithm A comes very close to this bound.

Lemma 7.2 *For all L_n we have*

$$A(L_n) \leq H_{1/2} + \Delta/2 + 1/4.$$

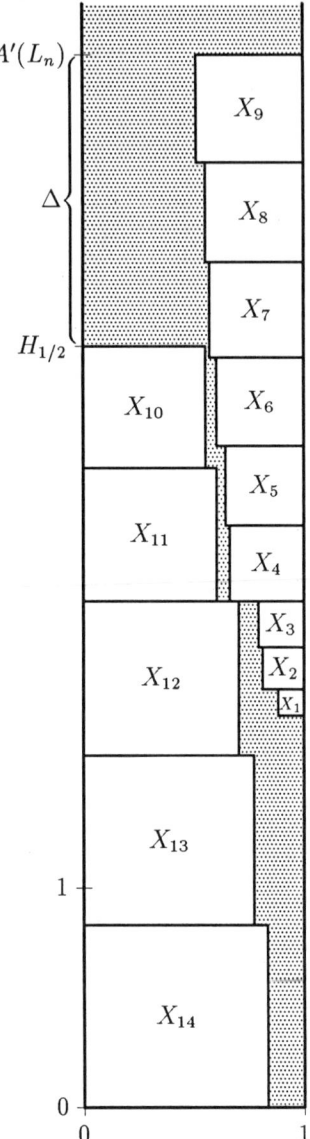

$L_n = (X_1, X_2, \ldots, X_{14})$
$= (0.12, 0.19, 0.21, 0.34, 0.36, 0.40, 0.43, 0.45, 0.49, 0.55, 0.60, 0.70, 0.77, 0.83)$

Figure 7.3: Example for the first three steps of Algorithm A.

7.1. OFF-LINE ALGORITHMS

Proof. Let A' denote the algorithm consisting of the first three steps of Algorithm A. Then $A'(L_n) = H_{1/2} + \Delta$ (see Figure 7.3). If $\Delta = 0$, we are done. Otherwise, the average height of the left and right stacks is $H_{1/2} + \Delta/2$; this average is not changed by the repacking in Step 4. After that repacking, we know that the stacks must be within $1/2$ of each other in height. Thus the maximum height is at most $1/4$ more than the average height, so we obtain the desired bound. ∎

Combining Lemmas 7.1 and 7.2, we see that $A(L_n) - \text{OPT}(L_n) \leq 1/4$ for all L_n such that $X^* > 1/3$. Clearly, Algorithm A is designed to do well only when all squares packed above $H_{1/2}$ exceed $1/3$, since such squares must be limited to two stacks. The main theorem below shows that the fraction of instances L_n for which this property does not hold tends to 0 very fast as $n \to \infty$.

Theorem 7.3 ([CL89]) *There is a constant $c > 0$ such that $\Pr\{X^* \leq 1/3\} = O(e^{-cn})$, and, consequently,*

$$\Pr\{A(L_n) - \text{OPT}(L_n) \leq 1/4\} = 1 - O(e^{-cn}).$$

Furthermore,

$$\mathsf{E}[A(L_n)] = \mathsf{E}[\text{OPT}(L_n)] + O(1).$$

Proof. Clearly, for the first part of the theorem it is enough to show that $\delta(y) \leq 0$ for all $1/6 \leq y \leq 1/2$ with probability $1 - O(e^{-cn})$ for some $c > 0$. In the analysis below it is convenient to view $\delta(y)$ as a random process on $[0, 1/2]$, as illustrated in Figure 7.4. The points (square sizes) selected in $[0, 1/2]$ are reflected about the midpoint $1/2$ onto the interval $[1/2, 1]$. These reflected points are labeled with a plus, whereas those falling originally in $(1/2, 1]$ are labeled minus. The pluses and minuses are then mapped by a simple translation onto $[0, 1/2]$. Sample functions for $\{\delta(y)\}_{0 < y \leq 1/2}$ are step functions, as illustrated in Figure 7.4. A square $X \in [0, 1/2]$ becomes a plus at $y = 1/2 - X$ and creates a positive step of size $X \leq 1/2$ in $\delta(y)$. A square $X \in (1/2, 1]$ becomes a minus at $y = X - 1/2$ and creates a negative step of size $X \geq 1/2$ in $\delta(y)$. The event locations comprise n independent uniform random draws from the interval $0 \leq y \leq 1/2$ in Figure 7.4 (i.e., the mappings $y = 1/2 - X$ for $X \leq 1/2$ and $y = X - 1/2$ for $X \geq 1/2$ produce uniform random draws from $[0, 1/2]$, since X is a uniform random draw from $[0, 1]$). The sign of each event is equally likely to be $+$ or $-$, independent of its location.

$L_n = (0.15, 0.22, 0.35, 0.45, 0.6, 0.7, 0.96)$

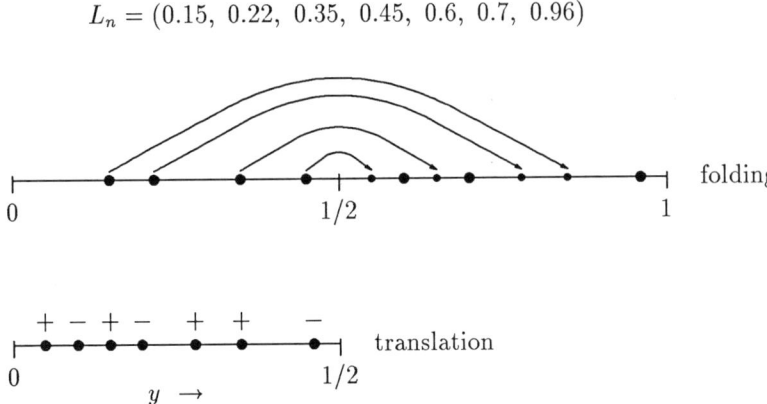

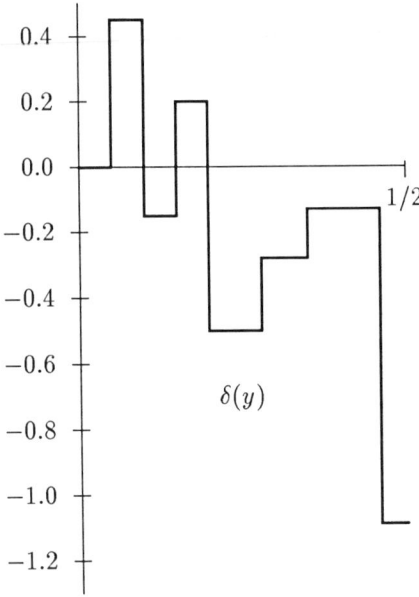

Figure 7.4: The function $\delta(y)$.

7.1. OFF-LINE ALGORITHMS

Now choose fixed α and β satisfying $0 < \alpha < 1/36$, $2/3 < \beta < 1$. Let $N(y)$ denote the number of events in the interval $[1/2 - y, 1/2 + y]$, and consider those samples L_n such that

$$\delta(1/6) < -\alpha n \quad \text{and} \quad N(1/6) \leq \beta n \qquad (7.5)$$

(We verify later that (7.5) holds with very high probability.) For convenience, assume that $1/3\alpha$ is an integer and that it divides $n - N(1/6)$, the number of squares in $[0, 1/3]$ plus the number in $(2/3, 1]$. It will be easy to check that these assumptions have no effect on our asymptotic results for large n.

For any list satisfying (7.5) divide up the at least $n - N(1/6)$ events of $\delta(y)$ at locations $1/6 \leq y \leq 1/2$ into $1/3\alpha$ blocks of $k = 3\alpha\bigl(n - N(1/6)\bigr) \leq 3\alpha n$ events each. Let δ_j denote the value of $\delta(y)$ just after the last event of block j, $1 \leq j \leq 1/3\alpha$, and let $\delta_0 = \delta(1/6)$. The sizes of positive steps are at most $1/3$ and are decreasing as y varies from $1/6$ to $1/2$. Therefore, no block of at most $3\alpha n$ events starting at $\delta_{j-1} < -\alpha n$ can include a zero crossing, where $\delta(y)$ would become positive. Thus for $\delta(y)$ to become positive in $1/6 \leq y \leq 1/2$ there must be at least one $j \geq 1$ such that $\delta_{j-1} < -\alpha n$ and $\delta_j \geq -\alpha n$. But at any y in $[1/6, 1/2]$ the magnitude $(1/2 + y)$ of a negative jump is at least twice the magnitude $(1/2 - y)$ of a positive jump. Hence, for any $j \geq 1$, the final value δ_j can be greater than the starting value δ_{j-1} only if the number of pluses, N_j^+, was greater than twice the number of minuses, N_j^-, in the jth block. Hence, if $\delta(1/6) < -\alpha n$ and $N_j^+ \leq 2N_j^-$ for all $j \geq 1$, then $\delta(y) \leq 0$ for $1/6 \leq y \leq 1/2$.

If $p_{\alpha\beta}$ denotes the probability that an arbitrary L_n satisfies both inequalities in (7.5), then by the above argument and Boole's inequality

$$\Pr\{X^* \leq 1/3\} \leq \Pr\{\delta(y) \leq 0 \text{ for all } y \in [1/6, 1/2]\}$$

$$\leq (1 - p_{\alpha\beta}) + \sum_{j=1}^{1/3\alpha} \Pr\{N_j^+ > 2N_j^-\}$$

$$\leq (1 - p_{\alpha\beta}) + \frac{1}{3\alpha} \max_j \Pr\{N_j^+ > 2N_j^-\}. \qquad (7.6)$$

To bound the term $1 - p_{\alpha\beta}$ in (7.6), first express $\delta(y)$ as a sum of i.i.d. random variables (see (7.3)), $\delta(y) = \sum_{i=1}^n Z_i$, where $|Z_i| \leq 1$, $1 \leq i \leq n$, and

$$Z_i = \begin{cases} X_i & \text{if } X_i \in [1/2 - y, 1/2], \\ -X_i & \text{if } X_i \in [1/2, 1/2 + y], \\ 0 & \text{otherwise.} \end{cases} \qquad (7.7)$$

A simple calculation yields

$$E[Z_i] = -y^2, \quad E[\delta(y)] = -y^2 n, \tag{7.8}$$

so that $E[\delta(1/6)] = -n/36$ and

$$\Pr\{\delta(1/6) > -\alpha n\} = \Pr\{\delta(1/6) - E[\delta(1/6)] > (1/36 - \alpha)n\}.$$

Then, by the Hoeffding bound of Theorem 2.5 (page 19) and our assumption $\alpha < 1/36$, there exists a $c > 0$ such that

$$\Pr\{\delta(1/6) > -\alpha n\} < e^{-cn}. \tag{7.9}$$

Similarly, we can write $N(y) = \sum_{i=1}^n Y_i$ with $|Y_i| \le 1$, $1 \le i \le n$, where

$$Y_i = \begin{cases} 1 & \text{if } X_i \in [1/2 - y, 1/2 + y], \\ 0 & \text{otherwise}, \end{cases} \tag{7.10}$$

and

$$E[Y_i] = 2y, \quad E[N(y)] = 2yn.$$

Then $E[N(1/6)] = n/3$ and $\beta > 2/3$ imply

$$\Pr\{N(1/6) > \beta n\} = \Pr\{N(1/6) - n/3 > (\beta - 1/3)n\}$$
$$< \Pr\{N(1/6) - n/3 > n/3\},$$

whereupon the Hoeffding bound yields

$$\Pr\{N(1/6) > \beta n\} < e^{-n/18}. \tag{7.11}$$

Applying Boole's inequality with (7.9) and (7.11), we have the desired estimate for the first term in (7.6): there exists a $c > 0$ such that

$$1 - p_{\alpha\beta} = O(e^{-cn}). \tag{7.12}$$

For the second term in (7.6), observe that there will be more than twice as many pluses as minuses in a block if and only if the number of pluses exceeds $2/3$ the total of k signs. Then

$$\Pr\{N_j^+ > 2N_j^-\} = \Pr\{N_j^+ - k/2 > k/6\} \quad \text{for } j \ge 1. \tag{7.13}$$

Note that N_j^+ can be expressed as a sum of k 0-1 random variables, so the bound of Theorem 2.5 can again be used. Since $k \ge 3\alpha(1-\beta)n - 1$, this bound and $\beta < 1$ imply that there exists a $c > 0$ such that

$$\Pr\{N_j^+ > 2N_j^-\} \le O(e^{-cn}). \tag{7.14}$$

7.1. OFF-LINE ALGORITHMS

Substituting (7.12) and (7.14) into (7.6), we obtain as desired a $c > 0$ such that $\Pr\{X^* \leq 1/3\} = O(e^{-cn})$.

The expected-height result is easy. We need only use the trivial bound $A(L_n) \leq n$ in order to write, by the use of a bound that usually holds,

$$\mathsf{E}[A(L_n) - \text{OPT}(L_n)] \leq \frac{1}{4}\left(1 - O(e^{-cn})\right) + O(ne^{-cn}) = O(1). \qquad \blacksquare$$

An asymptotic estimate for large n can also be determined for the expected difference between $A(L_n)$ and the lower bound $3n/8$; by Theorem 7.3 such a result and $\mathsf{E}[\text{OPT}(L_n) - 3n/8]$ must agree within an additive constant. By applying standard limit theorems, it is not difficult to discover the form such a result is likely to take. The following heuristic argument leads only to an educated guess, but as such it illustrates the original conception of a result, which is often obscured by the details of a rigorous proof.

Expressing $\Delta_n = \Delta(L_n)$ as an explicit function of n, it is clear from Lemma 7.1, Lemma 7.2, and Theorem 7.3 that $\mathsf{E}[A(L_n) - 3n/8] = \Theta(\mathsf{E}[\Delta_n])$. By the definition of $\Delta(L_n)$, we seek a largest function of n, say $f(n)$, such that for some sequence $y_1, y_2, \ldots$ we have $\Pr\{\delta(y_n) > f(n)\}$ bounded away from 0 for all n sufficiently large. Then $\mathsf{E}[\delta(y_n)] = \Theta(f(n))$ will be taken as the proposed result for $\mathsf{E}[A(L_n) - 3n/8]$.

Consider any fixed y and recall that $\delta(y)$ can be expressed as the sum of i.i.d. random variables Z_i as given in (7.7), with the mean

$$\mathsf{E}[\delta(y)] = -y^2 n \qquad (7.15)$$

and standard deviation

$$\sigma[\delta(y)] = \sqrt{n(y/2 + 2y^3/3 - y^4)}. \qquad (7.16)$$

Therefore, for large n, $\delta(y)$ is approximately normally distributed. Now suppose that the normal limit law also applies when y varies according to the desired sequence $\{y_n\}$; i.e., for any fixed x,

$$\Pr\left\{\frac{\delta(y_n) - \mathsf{E}[\delta(y_n)]}{\sigma[\delta(y_n)]} > x\right\} \sim 1 - \Phi(x) > 0 \quad \text{as } n \to \infty. \qquad (7.17)$$

Then we want the sequence $\{y_n\}$ to yield a largest $f(n)$ such that for some constant $c > 0$

$$\frac{f(n) - \mathsf{E}[\delta(y_n)]}{\sigma[\delta(y_n)]} = \frac{f(n) + n y_n^2}{\sqrt{n(y_n/2 + 2y_n^3/3 - y_n^4)}} \sim c \quad \text{as } n \to \infty.$$

Ignoring lower-order terms in the denominator, we take $f(n) = \Theta(\sqrt{y_n n})$ and maximize y_n subject to the constraint $\sqrt{y_n n} = \Omega(\mathsf{E}[\delta(y_n)]) = \Omega(n y_n^2)$. This last constraint requires that $y_n = O(n^{-1/3})$, so we choose $y_n = n^{-1/3}$ and $f(n) = \Theta(n^{1/3})$. Then $\mathsf{E}[A(L_n) - 3n/8] = \Theta(n^{1/3})$, and hence

$$\mathsf{E}[\mathrm{OPT}(L_n)] - 3n/8 = \Theta(n^{1/3}) \tag{7.18}$$

becomes the proposed result.

The estimate (7.18) has, in fact, been proven rigorously in [CL89]; the details of the above approach are worked out in a proof of the lower bound $\Omega(n^{1/3})$, and a technique similar to Theorem 7.3 is used to prove the upper bound $O(n^{1/3})$.

7.1.2 Packing rectangles into a strip

The result corresponding to (7.18) in the case of rectangles is easier to derive, assuming that all dimensions are independent uniform random draws from $[0,1]$. In the remainder of this chapter this will be referred to as the *uniform model of rectangles*. A lower bound can be found by a simple generalization of the one-dimensional argument in Section 5.1. First, note that the height of the packing must be at least as large as the total area of the squares; it must also be at least as large as the total height of the squares with width exceeding $1/2$. Thus an analysis like that leading to Theorem 5.2 yields

$$\mathsf{E}[\mathrm{OPT}(L_n)] = \frac{n}{4} + \Omega(\sqrt{n}). \tag{7.19}$$

Next, return to Algorithm A and adapt it to rectangles as follows (Figure 7.5 illustrates the algorithm). Let $H_{1/2}$ denote the sum of the heights of the rectangles with widths exceeding $1/2$.

Algorithm B

1. Stack the rectangles with widths greater than $1/2$ along the left edge of the strip in order of decreasing width.

2. Starting at height $H_{1/2}$, stack the remaining squares along the right edge of the strip in order of increasing width.

3. Slide the stack on the right edge down until it rests on the bottom of the strip, or a rectangle in the right stack comes in contact with a rectangle in the left stack, whichever occurs first.

7.1. OFF-LINE ALGORITHMS

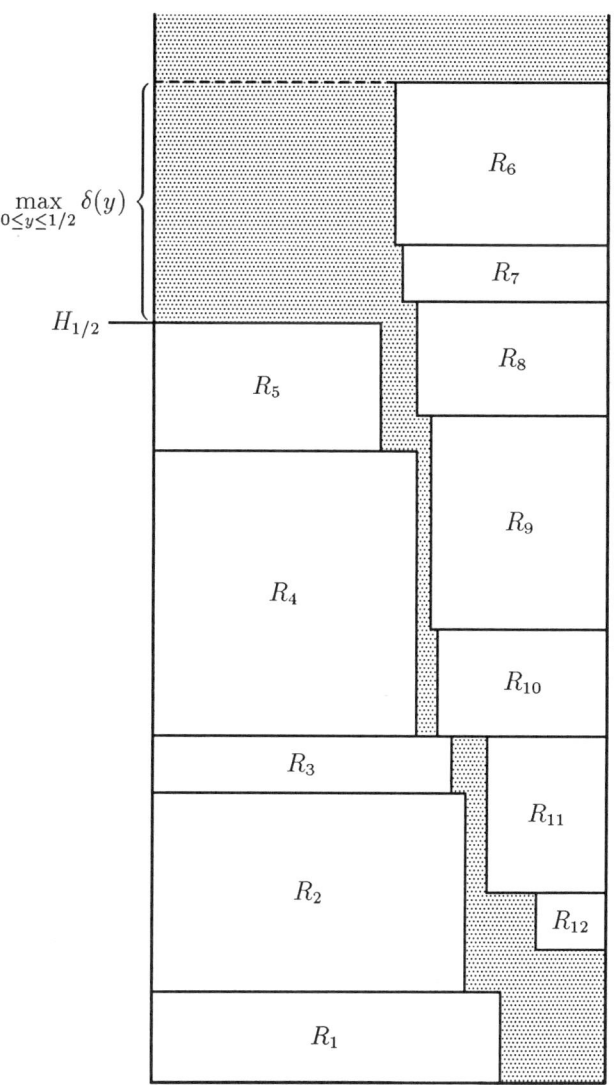

Figure 7.5: A packing produced by Algorithm B.

In analogy with the case of squares, define $\delta(0) = 0$ and

$$\delta(y) = \sum_{\frac{1}{2}-y \leq X_i \leq \frac{1}{2}} Y_i - \sum_{\frac{1}{2} < X_i \leq \frac{1}{2}+y} Y_i \quad \text{for } 0 < y \leq \frac{1}{2}. \tag{7.20}$$

It is clear from the definition of Algorithm B that

$$B(L_n) = H_{1/2} + \max_{0 \leq y \leq 1/2} \delta(y). \tag{7.21}$$

(See the example in Figure 7.5.) Here, the process $\delta(y)$ is easier to analyze than in the case of squares, for the step sizes are now i.i.d. random variables. Indeed, we may confine ourselves to the following random walk with a symmetric step distribution. In order of increasing y, let Z_i be the ith step in $\delta(y)$, and define the random walk $\xi_j = \sum_{i=1}^{j} Z_i$, $1 \leq j \leq n$, with $\xi_0 = 0$. Since the Y_i are independent, each with the distribution $U(0, 1)$, we have from (7.20) that the Z_i are independent uniform random draws from $[-1, 1]$. Clearly,

$$\max_{0 \leq j \leq n} \xi_j = \max_{0 \leq y \leq 1/2} \delta(y),$$

so from (7.21),

$$\mathsf{E}[B(L_n)] = \mathsf{E}[H_{1/2}] + \mathsf{E}\left[\max_{0 \leq j \leq n} \xi_j\right] = \frac{n}{4} + \mathsf{E}\left[\max_{0 \leq j \leq n} \xi_j\right].$$

Now

$$\mathsf{E}\left[\max_{0 \leq j \leq n} \xi_j\right] = \int_0^\infty \Pr\left\{\max_{0 \leq j \leq n} \xi_j > x\right\} dx = \sqrt{n} \int_0^\infty \Pr\left\{\max_{0 \leq j \leq n} \xi_j > x\sqrt{n}\right\} dx.$$

The Z_i are i.i.d. and bounded by 1 in absolute magnitude, so Theorem 2.7 (page 20) applies to $\max_{0 \leq j \leq n} \xi_j$ and shows that the integral in (7.21) converges. In conjunction with (7.19) we therefore have the following tight bound on optimum packings.

Theorem 7.4 ([CSa]) *For a strip packing of n rectangles with each width and height i.i.d. from $U(0, 1)$, we have*

$$\mathsf{E}[\mathrm{OPT}(L_n)] = \frac{n}{4} + \Theta(\sqrt{n}).$$

7.1.3 Two-dimensional bin packing

Consider the probabilistic analysis of algorithms for packing unit squares (bins) with a list L_n of rectangles described by the uniform model. Again, a lower bound can be found by a simple generalization of the one-dimensional argument in Section 5.1. The height of the packing must be at least as large as the total area of the squares; it must also be at least as large as the number of squares with both height and width exceeding 1/2, since clearly no two such items can fit together in a bin. Then an analysis like that leading to Theorem 5.2 yields the following result of [KLMS84]:

$$\mathrm{E}[\mathrm{OPT}(L_n)] = \frac{n}{4} + \Omega(\sqrt{n}). \tag{7.22}$$

(Note in fact that this argument holds even in the case in which we are allowed to rotate the items by 90°.)

We now define the extension of one-dimensional matching to two dimensions, which provides the basis for a two-dimensional bin-packing algorithm [KLMS84]. The algorithm first divides the items into the sets S_1, S_2, S_3, and S_4 containing items with dimensions in the quadrants of the unit square in the order $[0, 1/2]^2$, $[0, 1/2] \times [1/2, 1]$, $[1/2, 1] \times [0, 1/2]$, and $[1/2, 1]^2$, respectively. The algorithm then attempts to pack as many bins as possible with four items, one from each set. This is done by producing 3 one-dimensional (pairwise) matchings, M_{12}, M_{34}, and M_{24}, as follows. Let $I_i = (X_i, Y_i)$ denote the ith item. M_{12} provides a maximum matching of items in S_1 with those from S_2 such that for all $(I_i, I_j) \in M_{12}$ with $I_i \in S_1$ and $I_j \in S_2$ we have

$$Y_i + Y_j \leq 1 \quad \text{(the vertical dimensions fit)} \tag{7.23}$$

and

$$X_i \leq X_j \quad \text{(the taller item is at least as wide as the shorter item).} \tag{7.24}$$

Similarly, M_{34} is a maximum matching of items in S_3 and S_4 such that if $(I_i, I_j) \in M_{34}$ with $I_i \in S_3$ and $I_j \in S_4$, then (7.23) and (7.24) hold. Finally, M_{24} is a maximum matching of items in S_2 and S_4 such that if $(I_i, I_j) \in M_{24}$ with $I_i \in S_2$ and $I_j \in S_4$, then

$$X_i + X_j \leq 1 \quad \text{(the horizontal dimensions fit),} \tag{7.25}$$

i.e., M_{24} is a simple matching of the type described for MATCH in Section 5.1. A detailed example is shown in Figure 7.6.

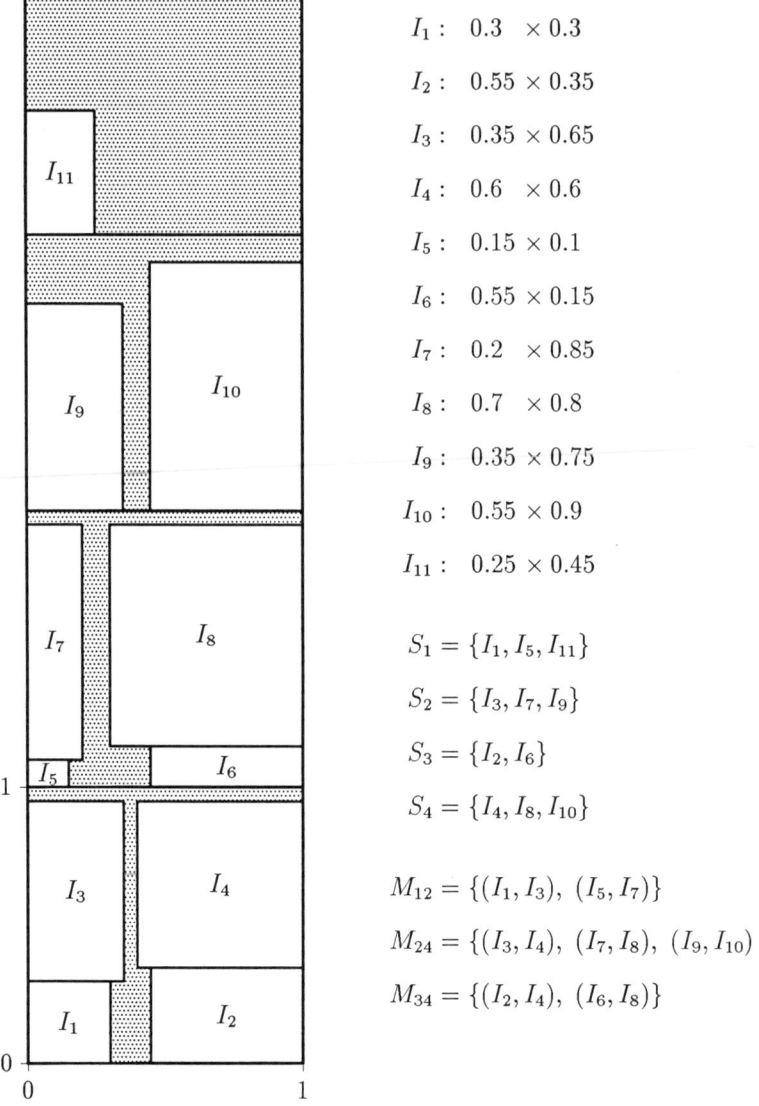

Figure 7.6: Example for Algorithm KLM.

7.1. OFF-LINE ALGORITHMS

For the analysis of this algorithm it is convenient to assume that the item sizes are generated by a Poisson process in two dimensions with rate parameter n (see Section 2.7.1). Thus, the number N of items is Poisson distributed with mean n and the (X_i, Y_i) are independent random points uniform over the unit square $[0,1]^2$. By this means we will be able to exploit the property that the numbers of points generated in disjoint regions of $[0,1]^2$ are independent.

Clearly, any pair of items matched in M_{12}, M_{24}, or M_{34} can be packed into a single bin. However, if $(I_i, I_j) \in M_{12}$ and $(I_k, I_\ell) \in M_{34}$, then by conditions (7.24) and (7.25) all four of these items can be placed into a bin if $(I_j, I_\ell) \in M_{24}$. Thus we define Algorithm KLM by the procedure that packs items I_i and I_j into the same bin if and only if either (I_i, I_j) is in one of the three matchings or there is a pair $(I_i', I_j') \in M_{24}$ such that $(I_i, I_i') \in M_{12}$ and $(I_j, I_j') \in M_{34}$; in this last case, the bin will contain the four items I_i, I_j, I_i', I_j' (e.g., in Figure 7.6, (I_1, I_2) is not in any matching but I_1 and I_2 occupy the same bin as I_3 and I_4, because (I_1, I_3), (I_2, I_4), and (I_3, I_4) are in M_{12}, M_{34}, and M_{24}, respectively).

Now let U_{12}, U_{34}, and U_{24} denote the numbers of items unmatched by M_{12}, M_{34}, and M_{24}, respectively. Since any bin with fewer than four items has at least one item counted in U_{12}, U_{34}, or U_{24}, we have the bound

$$\text{KLM}(L_n) \leq \frac{n}{4} + U_{12} + U_{34} + U_{24}.$$

Thus, since U_{12} and U_{34} are identically distributed,

$$\mathsf{E}[\text{KLM}(L_n)] \leq \frac{n}{4} + 2\mathsf{E}[U_{12}] + \mathsf{E}[U_{24}].$$

By the analysis in Section 5.1 we have $\mathsf{E}[U_{24}] = \Theta(\sqrt{n})$, so it remains to find the asymptotics of $\mathsf{E}[U_{12}]$.

For this analysis consider the square $[0, 1/2]^2$. For each I_i in S_1 we place a point (X_i, Y_i) in $[0, 1/2]^2$ and indicate it by a minus. Next, for each item I_j in S_2 we place a point $(X_j, 1 - Y_j)$ and indicate it by a plus. Now observe that in $[0, 1/2]^2$ a minus (X, Y) and a plus (X', Y') correspond to items in S_1 and S_2 that can be selected for M_{12} if and only if the plus is to the right of the minus (i.e., $X \leq X'$ so that the item in S_2 is at least as wide as the one in S_1) and above the minus (i.e., $Y' \geq Y$ or $(1 - Y') + Y \leq 1$, so that the vertical dimensions of the corresponding items sum to no more than 1).

It remains to observe that $1 - Y$ is uniformly distributed over $[0, 1/2]$, since Y is uniform over $[1/2, 1]$. It follows easily that, with suitable rescaling, a random instance of the matching problem producing M_{12} is a random instance of maximum up-right matching (see Section 3.2). By Theorem 3.2 (page 43) we have $E[U_{12}] = \Theta(\sqrt{n} \log^{3/4} n)$, and hence the following result.

Theorem 7.5 ([KLMS84]) *Let n rectangles be drawn according to the uniform model. Then*

$$E[\text{KLM}(L_n)] = \frac{n}{4} + \Theta(\sqrt{n} \log^{3/4} n). \qquad (7.26)$$

(Our analysis leading to this theorem follows that of [KLMS84], except that we use the stronger bounds which have since been proven on up-right matching to state a stronger result.)

The discrepancy between (7.22) and (7.26) shows that there is room for improvement in one or the other.

7.2 On-line algorithms

In an early paper on the average-case analysis of strip packing, Hofri [Hofr80] analyzed an extension of the NF rule to two dimensions. In packing L_n, this *Next Fit Level* (NFL) algorithm starts out by placing items left-justified along level 1, i.e., the bottom of the strip. If an item I_i is encountered that is too wide to fit in the remaining space, I_i is placed left-justified on a horizontal baseline drawn through the top of the tallest rectangle on level 1. Items $I_{i+1}, I_{i+2}, \ldots$ are then placed on this new level, level 2, until an item I_j, $j \geq i+1$, is found not to fit. Level 3 with I_j as its first item is then established on a baseline through the tallest item on level 2; this process continues until all items are packed.

Hofri also studies two variants: *Rotatable Next Fit* (RNF) and NF Decreasing Height (NFDH). RNF always places the larger of the two dimensions along the baseline, rotating items 90° where necessary. NFDH is the same as NFL except that L is assumed to be ordered by decreasing height; thus NFDH is not on-line.

Noting the correspondence between the levels of NFL strip packings and the bins of one-dimensional NF bin packing, we see that for NFL and NFDH the random variables W_i (the width of the first item in level i), N_i (the number of items in level i), and ℓ_i (the occupancy on the baseline of level i) all follow the same probability laws derived for the corresponding quantities in the one-dimensional problem. In making use of this observation, the analysis of NFL and NFDH strip packing follows lines similar to those in [CSHY80, Hofr84]. Of particular interest is the random variable W with the limiting distribution of the unused space W_i in level i as $i \to \infty$.

7.2. ON-LINE ALGORITHMS

Consider the behavior of NFL under the uniform model of rectangles. Let $I_{i1}, I_{i2}, \ldots, I_{iN_i}$ denote the items in level $i \geq 1$, where $I_{ij} = (X_{ij}, Y_{ij})$, and where $N_i \geq 1$ is the number of items on level i. The strip width is 1 and the height of the level is $\max_{1 \leq j \leq N_i} Y_{ij}$, so

$$W_i = \max_{1 \leq j \leq N_i} Y_{ij} - \sum_{j=1}^{N_i} X_{ij} Y_{ij}.$$

If the two dimensions of a rectangle are independent, then

$$\mathsf{E}[W_i] = \mathsf{E}\left[\max_{1 \leq j \leq N_i} Y_{ij}\right] - \mathsf{E}[Y]\mathsf{E}[l_i],$$

where $\mathsf{E}[Y]$ is the mean height of the rectangles in L_n, and

$$l_i = \sum_{j=1}^{N_i} X_{ij}$$

corresponds to the bin level in the one-dimensional case. Now Theorem 6.9 shows that $\mathsf{E}[l_i] \sim 3/4$, and hence $\mathsf{E}[Y]\mathsf{E}[l_i] \sim 3/8$ as $i \to \infty$. Given N_i, we have that $\max_{1 \leq j \leq N_i} Y_{ij}$ is the largest of N_i independent uniform random draws from $[0,1]$. Given $N_i = m$, the expectation is $m/(m+1)$, so

$$\mathsf{E}[W] = \sum_{m \geq 1} \frac{m}{m+1} \Pr\{N = m\} - \frac{3}{8}, \tag{7.27}$$

where N has the stationary distribution of the N_i given in (6.13), namely, $q_m = 3(m^2 + 3m + 1)/(m+3)!$, $m \geq 1$, with the generating function

$$N(z) = \sum_{m \geq 1} \Pr\{N = m\} z^m = \frac{3}{z^3}\left(e^z(z-1)^2 - \left(1 - z - \frac{z^2}{2} + \frac{z^3}{6}\right)\right). \tag{7.28}$$

Routine manipulations show that the first term on the right of (7.27) is $1 - \int_0^1 N(z)\,dz$, so

$$\mathsf{E}[W] = \frac{5}{8} - \int_0^1 N(z)\,dz. \tag{7.29}$$

From (7.28), $N(z)\,dz$ can be rewritten as

$$N(z)\,dz = dK(z) - \frac{3}{2}\frac{e^z - 1}{z}\,dz,$$

where

$$K(z) = \frac{3}{2}\frac{1 + z - e^z}{z^2} - \frac{9}{2}\frac{1 - e^z}{z} - \frac{z}{2}.$$

Substitution into (7.29) leads to

$$E[W] = \frac{51}{8} - 3e + \frac{3}{2}\int_0^1 \frac{e^z - 1}{z} dz.$$

By [AS70, formula (5.1.40)] the integral is equal to $Ei(1) - \gamma$, where $Ei(1)$ is the exponential integral function evaluated at 1 (from tables in [AS70] this is $1.8951\ldots$), and where $\gamma = 0.5772\ldots$ is Euler's constant. Substitutions produce

$$E[W] = 0.19701\ldots. \tag{7.30}$$

By Theorem 6.9 the expected number of levels is asymptotically $2n/3$; the expected level height is $E[W] + 3/8 = 0.57201\ldots$ from (7.27), so

$$E[\text{NFL}(L_n)] \sim (0.38134\ldots)n.$$

Note that the expected occupied space $(n/4)$ is about $2/3$ of the expected space used.

NFL is just one algorithm in the class of level algorithms, i.e., algorithms that pack rectangles on levels, where level 1 is the bottom of the strip and level i, $i \geq 2$, is a horizontal baseline drawn through the top of the tallest rectangle on level $i-1$. The First Fit Level algorithm figures to be substantially more efficient than NFL, although it gives up the linear-time property. According to this rule, the next rectangle is packed on the lowest level where it will fit, creating a new level when necessary; only the width needs to fit for a rectangle to be placed on the currently highest level, but both the width and height must fit for it to be placed on any other current level. The one-dimensional Best Fit rule can also be extended in obvious ways in order to define a Best Fit Level algorithm.

Level algorithms are especially appropriate when there is no information available about the rectangles to be packed, and the packing must be on-line. When size distributions and the numbers of items to be packed are known in advance, asymptotically optimum algorithms are easily constructed from one-dimensional analogues. For example, with height distributions known in advance, the *shelf algorithms* can eliminate much of the space wasted vertically between items. A class of such algorithms is defined by a set of shelf heights, $0 < s_1 < s_2 < \cdots < s_k$, which are essentially preset levels. Rectangles with heights in $(s_{i-1}, s_i]$ are packed left-justified on shelves of height s_i, $1 \leq i \leq k$, where $s_0 \equiv 0$. Within each category of shelves, any one-dimensional bin-packing algorithm can be adopted to control the gaps (space wasted horizontally) in the shelves; each shelf can be interpreted as a bin.

7.2. ON-LINE ALGORITHMS

For example, a linear-time algorithm is provided by the Next Fit Shelf algorithm (see [BVZ89]). As in the one-dimensional case, the Best Fit Shelf (BFS) algorithm is much more efficient asymptotically. An example is pictured in Figure 7.7. By exploiting one-dimensional results, the asymptotics of BFS are easily found for the width (length) distributions assumed in the one-dimensional case. The analysis is illustrated below in the uniform model of rectangles.

Let the shelf sizes be $s_i = i/k$, $1 \leq i \leq k$, and let n_i be the number of rectangles in L_n with heights in $(s_{i-1}, s_i]$. For each i, n_i is binomially distributed with parameters $1/k$ and n, and with mean $\mathsf{E}[n_i] = n/k$. To obtain the expected number $\mathsf{E}[S_i]$ of shelves of height i/k in the BFS packing of L_n, we merely use the one-dimensional BF result (Section 6.2) in terms of the mean n/k; i.e., there exists an $\alpha > 0$, independent of both k and n, such that

$$\mathsf{E}[S_i] \leq \frac{n}{2k} + \alpha \sqrt{n/k} \log^{3/4}(n/k), \quad \text{for } 1 \leq i \leq k, \qquad (7.31)$$

for all n sufficiently large. Here, we have made use of the fact that the asymptotic result for up-right matching still holds for the binomial distribution governing the number of items. From (7.31) we find for n large enough

$$\mathsf{E}[\mathrm{BFS}(L_n)] \leq \sum_{i=1}^{k} \frac{i}{k} \mathsf{E}[S_i] \leq \frac{n}{4} \left(1 + \frac{1}{k}\right) + \frac{\alpha}{2}(k+1)\sqrt{n/k} \log^{3/4}(n/k). \quad (7.32)$$

Along with the corresponding lower bound supplied by the upright matching results, (7.32) leads to the following result.

Theorem 7.6 ([CSa]) *Let rectangles be described as in the uniform model. If $k = \lfloor n^{1/3} \rfloor$ shelf sizes for BFS are chosen to be $s_i = i/\lfloor n^{1/3} \rfloor$, $1 \leq i \leq \lfloor n^{1/3} \rfloor$, then*

$$\mathsf{E}[\mathrm{BFS}(L_n)] = \frac{n}{4} + \Theta(n^{2/3} \log^{3/4} n).$$

We remark that the shelf algorithms in the uniform model of rectangles are easily adapted to on-line two-dimensional bin-packing. For k odd we simply assign shelves of heights i/k and $1 - i/k$ to the same bins. The above analysis is easily modified to yield the same result as in Theorem 7.6 (only the constants hidden by the Θ-notation change).

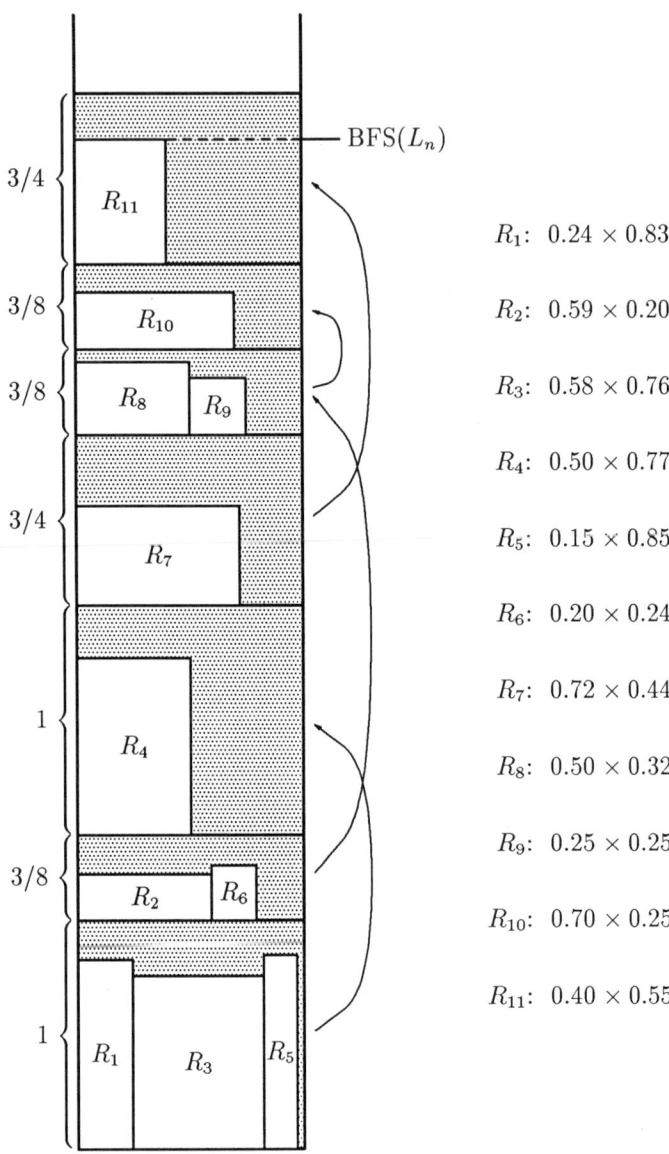

Figure 7.7: A BFS packing.

References

Each bibliography entry is followed by a list, in square brackets, of the pages on which it is referenced.

[AKT84] M. Ajtai, J. Komlós, and G. Tusnády. On optimal matchings. *Combinatorica*, 4:259–264, 1984. [41, 44, 45, 51, 56]

[AS70] M. Abramowitz and I. A. Stegun, editors. *Handbook of Mathematical Functions With Formulas, Graphs, and Mathematical Tables*. Applied Mathematics Series 55. National Bureau of Standards, Washington, D.C., 1970. [11, 12, 83, 93, 174]

[BC81] B. S. Baker and E. G. Coffman, Jr. A tight asymptotic bound for next-fit decreasing bin-packing. *SIAM Journal on Algebraic and Discrete Methods*, 2:147–152, 1981. [9]

[BD85] J. L. Bruno and P. J. Downey. Probabilistic bounds for dual bin packing. *Acta Informatica*, 22:333–345, 1985. [35]

[BD86] J. L. Bruno and P. J. Downey. Probabilistic bounds on the performance of list scheduling. *SIAM Journal on Computing*, 15:409–417, 1986. [35]

[Berg91] B. Berger. The fourth moment method. In *Proceedings of the Second Annual ACM-SIAM Symposium on Discrete Algorithms*, pages 373–383, 1991. [37]

[Bill65] P. Billard. Séries de fourier aléatoirement bornées, continues, uniformémente convergentes. *Annales Scientifiques Ecole Normale Superieure*, 82:131–179, 1965. [37]

[BJLM83] J. L. Bentley, D. S. Johnson, F. T. Leighton, and C. C. McGeoch. An experimental study of bin packing. In *Proceedings of the 21st Annual Allerton Conference on Communication,*

Control, and Computing, pages 51–60. University of Illinois, Urbana, 1983. [12, 126, 127]

[BJLMM84] J. L. Bentley, D. S. Johnson, F. T. Leighton, C. C. McGeoch, and L. A. McGeoch. Some unexpected expected behavior results for bin packing. In *Proceedings of the Sixteenth Annual ACM Symposium on Theory of Computing*, pages 279–288, 1984. [122–124, 127]

[BM76] J. A. Bondy and U. S. R. Murty. *Graph Theory with Applications*. North-Holland, New York, 1976. [57]

[Boxm84] O. J. Boxma. A probabilistic analysis of the LPT scheduling rule. In E. Gelenbe, editor, *Performance '84: Proceedings of the Tenth International Symposium on Models of Computer System Performance*, pages 475–490, Paris, France, December 1984. North-Holland. [80, 83]

[Boxm85] O. J. Boxma. A probabilistic analysis of multiprocessor list scheduling: the Erlang case. *Stochastic Models*, 1:209–220, 1985. [26]

[Brei68] L. Breiman. *Probability*. Addison-Wesley Series in Statistics. Addison-Wesley, Reading, MA, 1968. [20]

[BT89] K. H. Borgwardt and B. Tremel. The average quality of greedy algorithms for the subset-sum-maximization problem. Preprint number 198, Institut für Mathematik, University of Augsburg, 1989. [121]

[BVZ89] J. J. Bartholdi, J. H. Vande Vate, and J. Zhang. Expected performance of the shelf heuristic for two-dimensional packing. *Operations Research Letters*, 8:11–16, 1989. [175]

[CCF84] A. R. Calderbank, E. G. Coffman, Jr., and L. Flatto. Optimum head separation in a disk system with two read/write heads. *Journal of the ACM*, 31(4):826–838, October 1984. [138]

[CCGJM&91] E. G. Coffman, Jr., C. A. Courcoubetis, M. R. Garey, D. S. Johnson, L. A. McGeogh, P. W. Shor, R. R. Weber, and M. Yannakakis. Average-case performance of one-dimensional bin packing algorithms under discrete uniform distributions.

In *Proceedings of the 23rd Annual ACM Symposium on Theory of Computing*, 1991. To appear. [132]

[CFFGR86] J. Csirik, J. B. G. Frenk, A. Frieze, G. Galambos, and A. H. G. Rinnooy Kan. A probabilistic analysis of the next fit decreasing bin packing heuristic. *Operations Research Letters*, 5:233–236, 1986. [27]

[CFGR] J. Csirik, J. B. G. Frenk, G. Galambos, and A. H. G. Rinnooy Kan. Probabilistic analysis of algorithms for dual bin packing problems. *Journal of Algorithms*. To appear. [39, 101, 103]

[CFJR90] E. G. Coffman, Jr., G. Fayolle, P. Jacquet, and P. Robert. Largest-first sequential selection with a sum constraint. *Operations Research Letters*, 9:141–146, 1990. [121]

[CFL84a] E. G. Coffman, Jr., L. Flatto, and G. S. Lueker. Expected makespans for largest-first multiprocessor scheduling. In E. Gelenbe, editor, *Performance '84: Proceedings of the Tenth International Symposium on Models of Computer System Performance*, pages 475–490, Paris, France, December 1984. North-Holland. [77, 81]

[CFL84b] E. G. Coffman, Jr., G. N. Frederickson, and G. S. Lueker. A note on expected makespans for largest-first sequences of independent tasks on two processors. *Mathematics of Operations Research*, 9(2):260–266, May 1984. [33]

[CFW87] E. G. Coffman, Jr., L. Flatto, and R. R. Weber. Optimal selection of stochastic intervals under a sum constraint. *Journal of Applied Probability*, 19:454–473, 1987. [121]

[CG85] E. G. Coffman, Jr. and E. N. Gilbert. On the expected relative performance of list scheduling. *Operations Research*, 33:548–561, 1985. [26]

[CG86] J. Csirik and G. Galambos. An $O(n)$ bin-packing algorithm for uniformly distributed data. *Computing*, 36:313–319, 1986. [148]

[CGJ84] E. G. Coffman, Jr., M. R. Garey, and D. S. Johnson. Approximation algorithms for bin-packing—an updated survey. In G. Ausiello, M. Lucertini, and P. Serafini, editors, *Algorithm*

Design for Computer System Design, pages 49–106, Wein, 1984. Springer-Verlag. CISM Courses and Lectures Number 284. [3]

[Cher52] H. Chernoff. A measure of asymptotic efficiency for tests of a hypothesis based on the sum of observations. *The Annals of Mathematical Statistics*, 23:493–507, 1952. [16]

[Çinl75] E. Çinlar. *Introduction to Stochastic Processes*. Prentice-Hall, Englewood Cliffs, NJ, 1975. [32]

[CKWW89] C. Courcoubetis, P. Konstantopoulous, J. Walrand, and R. R. Weber. Stabilizing an uncertain production system. *Queueing Systems: Theory and Applications*, 5(1-3):37–54, 1989. [120]

[CL89] E. G. Coffman, Jr. and J. C. Lagarias. A probabilistic analysis of square packing. *SIAM Journal on Computing*, 18:166–185, 1989. [161, 166]

[CLR88] E. G. Coffman, Jr., G. S. Lueker, and A. H. G. Rinnooy Kan. Asymptotic methods in the probabilistic analysis of sequencing and packing heuristics. *Management Science*, 34(3):266–290, March 1988. Focussed Issue on Heuristics. [viii]

[Coff76] E. G. Coffman, Jr. *Computer and Job-Shop Scheduling Theory*. John Wiley & Sons, 1976. [7]

[Coff82] E. G. Coffman, Jr. An introduction to proof techniques for bin-packing approximation algorithms. In M. A. H. Dempster, J. K. Lenstra, and A. H. G. Rinnooy Kan, editors, *Deterministic and Stochastic Scheduling*, pages 245–270, Dordrecht, Holland, 1982. D. Reidel Publishing Co. Proceedings of the NATO Advanced Study and Research Institute on Theoretical Approaches to Scheduling Problems, Durham, England, July 6–17, 1981. [107]

[Cox62] D. R. Cox. *Renewal Theory*. Methuen's Monographs on Applied Probability and Statistics. Methuen & Company Ltd., science paperback edition 1967 edition, 1962. Distributed in the USA by Barnes and Noble Inc. [39]

[CSa] E. G. Coffman, Jr. and P. W. Shor. Packing in two dimensions: Asymptotic average-case analysis of algorithms. *Algorithmica*. To appear. [153, 154, 168, 175]

REFERENCES

[CSb]   E. G. Coffman, Jr. and P. W. Shor. A simple proof of the $O(\sqrt{n}\log^{3/4} n)$ up-right matching bound. *SIAM Journal on Discrete Mathematics.* To appear. [56]

[CS90]   E. G. Coffman, Jr. and P. W. Shor. Average-case analysis of cutting and packing in two dimensions. *European Journal of Operational Research*, 44:134–144, 1990. [155]

[CSHY80]   E. G. Coffman, Jr., K. So, M. Hofri, and A. C. Yao. A stochastic model of bin-packing. *Information and Control*, 44(2):105–115, February 1980. [135, 136, 172]

[CW86a]   C. Courcoubetis and R. R. Weber. A bin-packing system for objects with sizes from a finite set: Necessary and sufficient conditions for stability and some applications. In *Proceedings of the 25th IEEE Conference on Decision and Control*, pages 1686–1691, Athens, Greece, December 1986. [32, 120]

[CW86b]   C. Courcoubetis and R. R. Weber. Necessary and sufficient conditions for stability of a bin packing system. *Journal of Applied Probability*, 23:989–999, 1986. [119]

[CW90]   C. Courcoubetis and R. R. Weber. Stability of on-line bin packing with random arrivals and long-run average constraints. *Probability in the Engineering and Information Sciences*, 1990. To appear. [120]

[Durb73]   J. Durbin. *Distribution Theory for Tests Based on the Sample Distribution Function.* Regional Conference Series in Applied Mathematics, 9. Society for Industrial and Applied Mathematics, Philadelphia, 1973. [34, 35]

[Dyck90]   H. Dyckhoff. A typology of cutting and packing problems. *European Journal of Operational Research*, 44:145–159, 1990. [3]

[ES74]   P. Erdős and J. Spencer. *Probabilistic Methods in Combinatorics.* Academic Press, New York, 1974. [37]

[Fell68]   W. Feller. *An Introduction to Probability Theory and Its Applications*, volume I. John Wiley & Sons, New York, third edition, 1968. [11, 14, 39, 40]

REFERENCES

[Fell71]   W. Feller. *An Introduction to Probability Theory and Its Applications*, volume II. John Wiley & Sons, New York, second edition, 1971. [11, 13, 14, 34, 92, 97]

[FF62]   L. R. Ford, Jr. and D. R. Fulkerson. *Flows in Networks*. Princeton University Press, 1962. [45]

[FK]   S. Floyd and R. M. Karp. FFD bin packing for item sizes with distributions on [0,1/2]. *Algorithmica*. To appear. An earlier version appeared as *Proceedings of the 27th Symposium on Foundations of Computer Science*, 1986, pp. 322–330. [124, 127]

[FR87]   J. B. G. Frenk and A. H. G. Rinnooy Kan. The asymptotic optimality of the LPT rule. *Mathematics of Operations Research*, 12(2):241–254, May 1987. [76, 83]

[Fred80]   G. N. Frederickson. Probabilistic analysis for simple one- and two-dimensional bin packing algorithms. *Information Processing Letters*, 11(4–5):156–161, December 1980. [99, 122]

[FV89]   D. P. Foster and R. V. Vohra. Probabilistic analysis of a heuristics for the dual bin packing problem. *Information Processing Letters*, 31:287–290, 1989. [121]

[GJ79]   M. R. Garey and D. S. Johnson. *Computers and Intractability: A Guide to the Theory of NP-Completeness*. W. H. Freeman, New York, 1979. [2]

[Glic78]   N. Glick. Breaking records and breaking boards. *American Mathematical Monthly*, 85:2–26, 1978. [121]

[Grah66]   R. L. Graham. Bounds for certain multiprocessing anomalies. *Bell System Technical Journal*, 45:1563–1581, 1966. [7]

[Grah69]   R. L. Graham. Bounds on multiprocessing timing anomalies. *SIAM Journal on Applied Mathematics*, 17:263–269, 1969. [8]

[HK86]   M. Hofri and S. Kamhi. A stochastic analysis of the NFD bin-packing algorithm. *Journal of Algorithms*, 7:489–509, 1986. [27]

REFERENCES

[Hoef63]   W. Hoeffding. Probability inequalities for sums of bounded random variables. *Journal of the American Statistical Association*, 58:13–30, 1963. [16, 19]

[Hoff82]   U. Hoffman. A class of simple stochastic online bin packing algorithms. *Computing*, 29:227–239, 1982. [148]

[Hofr80]   M. Hofri. Two dimensional packing: Expected performance of simple level algorithms. *Information and Control*, 45:1–17, 1980. [172]

[Hofr84]   M. Hofri. A probabilistic analysis of the next fit bin packing algorithm. *Journal of Algorithms*, 5:547–556, 1984. [172]

[Hofr87]   M. Hofri. *Probabilistic Analysis of Algorithms*. Springer-Verlag, New York, 1987. [11, 137]

[Jack41]   D. Jackson. *Fourier series and orthogonal polynomials*, volume 6 of *The Carus Mathematical Monographs*. The Mathematical Association of America, Oberlin, Ohio, 1941. [68]

[JDUGG74]   D. S. Johnson, A. Demers, J. D. Ullman, M. R. Garey, and R. L. Graham. Worst-case performance bounds for simple one-dimensional packing algorithms. *SIAM Journal on Computing*, 3(4):299–325, December 1974. [9]

[John]   D. S. Johnson. Private communication. [122]

[John73]   D. S. Johnson. *Near-Optimal Bin Packing Algorithms*. Ph.D. thesis, Massachusetts Institute of Technology, Department of Mathematics, Cambridge, 1973. [30]

[John74]   D. S. Johnson. Fast algorithms for bin packing. *Journal of Computer and System Sciences*, 8:272–314, 1974. [8]

[Karm82]   N. Karmarkar. Probabilistic analysis of some bin-packing algorithms. In *Proceedings of the 23rd Annual Symposium on Foundations of Computer Science*, pages 107–111, 1982. [104, 105, 137, 138, 144, 151]

[Karp72]   R. M. Karp. Reducibility among combinatorial problems. In R. E. Miller and J. W. Thatcher, editors, *Complexity of Computer Computations*, pages 85–103. Plenum Press, New York, 1972. [2]

[Karp82]   R. M. Karp. Lecture notes. Computer Science Division (EECS), University of California, Berkeley, 1982. [100, 105, 122]

[Karp85]   R. M. Karp. Private communication, 1985. [87]

[King76]   J. F. C. Kingman. *Subadditive Processes*, volume 539 of *Lecture Notes in Mathematics*, pages 167–223. Springer-Verlag, Berlin, 1976. Ecole d'Eté de Probabilitiés de Saint-Flour V— 1976. [146]

[KK82]   N. Karmarkar and R. M. Karp. The differencing method of set partitioning. Technical Report UCB/CSD 82/113, Computer Science Division (EECS), University of California, Berkeley, December 1982. [83, 86]

[KKLO86]   N. Karmarkar, R. M. Karp, G. S. Lueker, and A. M. Odlyzko. Probabilistic analysis of optimum partitioning. *Journal of Applied Probability*, 23(3):626–645, 1986. [90, 91, 96, 97]

[Klei76]   L. Kleinrock. *Queueing Systems*, volume I. John Wiley & Sons, New York, 1975-76. [127]

[KLMR84]   R. M. Karp, J. K. Lenstra, C. J. H. McDiarmid, and A. H. G. Rinnooy Kan. Probabilistic analysis of combinatorial algorithms. Technical Report OS-R8411, Centre for Mathematics and Computer Science, Amsterdam, 1984. [3]

[KLMS84]   R. M. Karp, M. Luby, and A. Marchetti-Spaccamela. A probabilistic analysis of multidimensional bin packing problems. In *Proceedings of the Sixteenth Annual ACM Symposium on Theory of Computing*, pages 289–298, 1984. [43, 100, 101, 130, 169, 172]

[KLV87]   K. Krause, L. Larmore, and D. Volper. Packing items from a triangular distribution. *Information Processing Letters*, 25:351–361, 1987. [28]

[Knöd81]   W. Knödel. *A Bin Packing Algorithm with Complexity $O(n \log n)$ and Performance 1 in the Stochastic Limit*, volume 118 of *Lecture Notes in Computer Science*, pages 369–378. Springer-Verlag, Berlin, 1981. *Mathematical Foundations of Computer Science 1981, Proceedings, 10th Symposium*, J.

Gruska and M. Chytil, eds., Štrbské Pleso, Czechoslovakia, August 31–September 4, 1981. [35, 100, 104, 105]

[Knut73] D. E. Knuth. *Fundamental Algorithms*, volume I of *The Art of Computer Programming*. Addison-Wesley, Reading, MA, second edition, 1973. [9, 15, 40, 93]

[Knut76] D. E. Knuth. Big omicron and big omega and big theta. *SIGACT News*, pages 18–24, April–June 1976. [6]

[KT81] S. Karlin and H. M. Taylor. *A Second Course in Stochastic Processes*. Academic Press, New York, 1981. [33]

[LL82] F. T. Leighton and C. E. Leiserson. Wafer-scale integration of systolic arrays. In *Proceedings of the 23rd Annual Symposium on Foundations of Computing*, pages 297–311, 1982. [43]

[LL87] C. C. Lee and D. T. Lee. Robust on-line bin packing algorithms. Technical report, Department of Electrical Engineering and Computer Science, Northwestern University, Evanston, IL, 1987. [147, 148]

[Loul84a] R. Loulou. Tight bounds and probabilistic analysis of two heuristics for parallel processor scheduling. *Mathematics of Operations Research*, 9:142–150, 1984. [75, 76]

[Loul84b] R. Loulou. Probabilistic behavior of optimal bin-packing solutions. *Operations Research Letters*, 3(3):129–135, August 1984. [105, 113]

[LS89] T. Leighton and P. Shor. Tight bounds for minimax grid matching with applications to the average case analysis of algorithms. *Combinatorica*, 9(2):161–187, 1989. [41, 56]

[Luek80] G. S. Lueker. Some techniques for solving recurrences. *ACM Computing Surveys*, 12(4):419–436, December 1980. [40]

[Luek82] G. S. Lueker. An average-case analysis of bin packing with uniformly distributed item sizes. Technical Report 181, University of California at Irvine, Department of Information and Computer Science, 1982. [100, 101, 122]

[Luek83] G. S. Lueker. Bin packing with items uniformly distributed over intervals $[a, b]$. In *Proceedings of the 24th Annual Symposium on Foundations of Computer Science*, pages 289–297, 1983. [28, 107]

[Luek87] G. S. Lueker. A note on the average-case behavior of a simple differencing method for partitioning. *Operations Research Letters*, 6(6):285–287, December 1987. [87, 90]

[Manb89] U. Manber. *Introduction to Algorithms: A Creative Approach*. Addison-Wesley, Reading, MA, 1989. [57]

[McGe87] C. C. McGeoch. *Experimental Analysis of Algorithms*. Ph.D. thesis, Carnegie-Mellon University, Department of Computer Science, 1987. [12]

[Murg88] F. D. Murgolo. Anomalous behavior in bin packing algorithms. *Discrete Applied Mathematics*, 21:229–243, 1988. [30, 128]

[OMW84] H. L. Ong, M. J. Magazine, and T. S. Wee. Probabilistic analysis of bin packing heuristics. *Operations Research*, 32:993–998, 1984. [12]

[PB85] P. W. Purdom, Jr. and C. A. Brown. *The Analysis of Algorithms*. Holt, Rinehart and Winston, New York, 1985. [40]

[Pipp77] N. Pippenger. An information-theoretic method in combinatorial theory. *Journal of Combinatorial Theory*, 23(1):99–104, July 1977. [95]

[Rama89] P. Ramanan. Average-case analysis of the smart next fit algorithm. *Information Processing Letters*, 31:221–225, June 12 1989. [151]

[RF86] A. H. G. Rinnooy Kan and J. B. G. Frenk. On the rate of convergence to optimality of the LPT rule. *Discrete Applied Mathematics*, 14:187–198, 1986. [76, 80, 83]

[Rhee87] W. T. Rhee. Probabilistic analysis of the next fit decreasing algorithm for bin packing. *Operations Research Letters*, 6(4):189–191, 1987. [27]

REFERENCES

[Rhee88]   W. T. Rhee. Optimal bin packing with items of random sizes. *Mathematics of Operations Research*, 13(1):140–151, February 1988. [110, 113, 114]

[Rhee90]   W. T. Rhee. A note on optimal bin packing and optimal bin covering with items of random size. *SIAM Journal on Computing*, 19(4):705–710, 1990. [36, 40]

[RT87]   W. T. Rhee and M. Talagrand. Martingale inequalities and NP-complete problems. *Mathematics of Operations Research*, 12(1):177–181, February 1987. [144, 146]

[RT88a]   W. T. Rhee and M. Talagrand. Exact bounds for the stochastic upward matching problem. *Transactions of the American Mathematical Society*, 307(1):109–125, May 1988. [41, 56]

[RT88b]   W. T. Rhee and M. Talagrand. Some distributions that allow perfect packing. *Journal of the ACM*, 35(3):564–578, July 1988. [28, 110, 114–117]

[RT89a]   P. Ramanan and K. Tsuga. Average-case analysis of the modified harmonic algorithm. *Algorithmica*, 4(4):519–533, 1989. [148]

[RT89b]   W. T. Rhee and M. Talagrand. The complete convergence of best fit decreasing. *SIAM Journal on Computing*, 18(5):909–918, 1989. [35, 128, 129]

[RT89c]   W. T. Rhee and M. Talagrand. The complete convergence of first fit decreasing. *SIAM Journal on Computing*, 18(5):919–938, 1989. [35, 128, 129]

[RT89d]   W. T. Rhee and M. Talagrand. Optimal bin covering with items of random size. *SIAM Journal on Computing*, 13(3):487–498, 1989. [40]

[RT89e]   W. T. Rhee and M. Talagrand. Optimal bin packing with items of random sizes—II. *SIAM Journal on Computing*, 18(1):139–151, 1989. [110, 113, 114]

[RT89f]   W. T. Rhee and M. Talagrand. Optimal bin packing with items of random sizes—III. *SIAM Journal on Computing*, 18(3):473–486, 1989. [36, 110]

[RT91]	W. T. Rhee and M. Talagrand. On line bin packing with items of random size. Manuscript, The Ohio State University, Columbus, OH, 1991. [133]
[Sche71]	M. Schechter. *Principles of Functional Analysis*. Academic Press, New York, 1971. [110, 111]
[Serf80]	R. Serfling. *Approximation Theorems of Mathematical Statistics*. Wiley, New York, 1980. [22, 76]
[Shep72a]	L. A. Shepp. Covering the circle with random arcs. *Israel J. Math.*, 11:328–345, 1972. [37]
[Shep72b]	L. A. Shepp. Covering the line with random intervals. *Zeitschrift für Wahrscheinlichkeitstheorie und verwandte Gebiete*, 23:163–170, 1972. [37]
[Shor85a]	P. W. Shor. Private communication, 1985. [117]
[Shor85b]	P. W. Shor. *Random Planar Matching and Bin Packing*. Ph.D. thesis, Massachusetts Institute of Technology, Cambridge, September 1985. [43, 44]
[Shor86]	P. W. Shor. The average-case analysis of some on-line algorithms for bin packing. *Combinatorica*, 6(2):179–200, 1986. [viii, 41, 53, 54, 56, 129, 130, 132]
[Shor90]	P. W. Shor. How to do better than best-fit: An improved on-line bin packing algorithm. Extended abstract, AT&T Bell Laboratories, Murray Hill, NJ 07974, 1990. [56, 133]
[Stan80]	T. A. Standish. *Data Structure Techniques*. Addison-Wesley, Reading, MA, 1980. [9]
[Stou74]	W. F. Stout. *Almost Sure Convergence*. Academic Press, New York, 1974. [145]
[Tala91]	M. Talagrand. Matching theorems and empirical discrepancy computations using majorizing measures. Manuscript, The Ohio State University, Columbus, OH, 1991. [51, 56]
[Tsai90]	L.-H. Tsai. Asymptotic analysis of an algorithm for balanced parallel-processor scheduling. Manuscript, 1990. Department of Decision and Information Sciences, University of Florida, Gainesville 32611. [90]

Index

Numbers in boldface indicate pages on which an index entry is defined.

absolute error, **75**
active bin, **8**
   limited number of, 150–154
average idle time, 75

Bernstein's bound, **19**, 64, 66
Berry-Esséen theorem, **13**
Best Fit, **8**, 30, 120, 129–132
Best Fit Decreasing, **9**, 30, 122–123, 128–129
Best Fit Level, **174**
Best Fit Shelf, **175**, 175
BF, see Best Fit
BFD, see Best Fit Decreasing
BFS, see Best Fit Shelf
bin covering, **39**, 39–40
bin packing, **1**
   applications, 4
   heuristics, 8–9, 26–30, 121–154
   optimum solution, 99–120
   two-dimensional, see two-dimensional bin packing
binomial distribution, 14–15, 23, 41, 66, 149, 175
Boole's inequality, **11**, 30, 164
bound that usually holds, 27–30, 68

Cauchy-Schwarz inequality, 11
centered partial sum, **13**
central limit theorem, 12–15

characteristic function, **13**, 92, 97
Chebyshev's inequality, 11, 37
Chernoff estimates, 15–16, 77
closed bin, **8**
closed-end algorithm, **7**
coin flips, 100
companion squares, **46**
conditioning, 3
configuration, **31**
Covering Next Fit, **39**

differencing method, 83–90
discrete distributions, 119
DM*, **86**, 86–87
dominating algorithm, **26**, 26–27, 122, 124, 129
dual-feasible function, **107**, 108, 110
   vs. subadditive function, 113

Edgeworth expansion, 14, 92
ellipsoid, volume of, 71
Euclidean matching, see matching problems, Euclidean
Euler's constant, 174
Euler's summation formula, 14
even partition, **90**
experimentation, 11, 107
exponential distribution
   and uniform, 32, 33, 77, 87, 100

190 INDEX

order statistics, 34, 87
exponential integral function Ei, 174

FF, *see* First Fit
FFD, *see* First Fit Decreasing
First Fit, **8**, 30, 132
First Fit Decreasing, **9**, 30, 122–129
First Fit Level, **174**
first-bin problem, **121**
Fourier analysis, 66
fourth moment method, **37**
functional, **111**
  linear, **111**
  sublinear, **111**

gamma function, 82
generating function, 40, 173
geometric-arithmetic mean inequality, 11, 19
greedy method, 8

Hahn-Banach theorem, **111**
  geometric form, 110
Hall's matching theorem, 57
HARMONIC algorithm, **146**, 146–148
Hoeffding bound, 16–21, 30, 164
Hölder's inequality, 37

indicator, 15
Interval First Fit, 148

Jensen's inequality, 11, 74, 106

k-conservative, **145**
Kolmogorov-Smirnov statistic, 34, 100, 128
  one-sided, 34

large deviations, 15–21

Largest Differencing Method, 84, 85
Largest Processing Time first, 8, 33, 75–83, 90
  rates of convergence, 76–83
law of large numbers, 36
LDM, *see* Largest Differencing Method
level algorithms, **174**
linear programming, 43, 45
linear programming relaxation, **31**, 36
  dual of, 106
linear-time algorithm, **8**, 121, 133–154, 174, 175
Lipschitz condition, 86
List Scheduling, **7**, 24, 25, 75
lower layer, 57
LPT, *see* Largest Processing Time first
LS, *see* List Scheduling

makespan scheduling, *see* scheduling
Markov chain, 23–25, 135–144, 172–174
martingale, **16**, 19, 144
MATCH algorithm, **100**, 122
matching problems, 41–74
  Euclidean, **41**, 43–53
  rightward, **43**, 53–56, 132–133
  total-edge-length, *see* matching problems, Euclidean
  up-right, **41**, 43, 56–74, 129–132, 172
MBF, *see* Modified Best Fit
median of a distribution, 81, 91–98
MFFD, *see* Modified First Fit Decreasing

INDEX

minimum total flow time, 33
Minkowski functional, 111
Modified Best Fit, **129**
Modified First Fit Decreasing, **124**
monotonic algorithm, **30**, 30–32, 123, 128, 129, 144
multiprocessor system, 1

Next Fit, **8**, 30, 133–146
    adapted to bin covering, *see* Covering Next Fit
Next Fit Decreasing, **9**, 26–27, 106
Next Fit Level, **172**
Next Fit Shelf, **175**
NF, *see* Next Fit
NF Decreasing Height, **172**
NFD, *see* Next Fit Decreasing
NFDH, *see* NF Descreasing Height
NFL, *see* Next Fit Level
normal distribution, **12**, 92, 102, 121, 165
normalized partial sum, **13**
notation for growth rates, **6**
$\mathcal{NP}$-completeness, 2

off-line algorithm, **7**, 9, 155–172
on-line algorithm, **7**, 8–9, 119, 129–154, 172–175
ONLINEMATCH, **148**, 148–150
open-end algorithm, **7**, 121, 132–133
optimum packing ratio, **104**
Optimum Restricted Completion, **122**
ORC, *see* Optimum Restricted Completion

packing constant, 35, 105, 110
Paired Differencing Method, **84**, 87–90

Paired Largest Processing Time first, **33**
Parseval's relation, 97
partial sum, **12**
partition problem, **1**, 75–98
    applications, 4
    heuristics, 24–26, 33–34, 75–90
    optimum solution, 90–98
PDM, *see* Paired Differencing Method
perfect packing, **104**, **105**, 104–120
PLPT, *see* Paired Largest Processing Time first
plus discrepancy, **45**
Poisson distribution, 32–34, 57
Poisson process, 32, 33, 77, 87, 171

queueing theory, 124–127

random walk, 20
    biased, 156
randomization, 63, 119
record-breaking problem, **121**
relative error, **25**
renewal theory, 38–40
resampling, 87
rightward matching, *see* matching problems, rightward
RNF, *see* Rotatable Next Fit
Rotatable Next Fit, **172**

sample distance, **128**
sample distribution function, **34**, 128
scheduling, **1**, *see* partition problem
Schwarz' inequality, 11, 37
second moment method, **37**, 93
shelf algorithms, **174**
Skorohod's inequality, 20
sliced NFD, 27

Smart Next Fit, **151**, 150–151
Smirnov estimate, 35
smoothness condition, 86
SNF, *see* Smart Next Fit
SNFD$_r$, *see* sliced NFD
spray paint, 115
Stirling's formula, 71
strip packing, 4, **155**, 155–168
subadditive function, 113
subadditive heuristic, **146**
sums of random variables, 12–23
swallow polynomials
    approach zero fast enough to, **14**, 94
symmetrization, **91**
symmetry independence, **44**, 48–50

Taylor's theorem, 19, 95
trim loss, **4**
two-dimensional bin packing, 5, 43, **155**, 169–172

$U(a,b)$, **6**
uniform distribution, 6
    order statistics, 32
uniform model of rectangles, **166**, 166–175
uniformly integrable, **22**
up-right matching, *see* matching problems, up-right

w-tight, **153**
weak convergence, **114**, 118, 119
weakly closed, **114**, 119
whp, *see* with high probability
with high probability, **64**